# Radiofrequency Heating of Plasmas

# The Adam Hilger Series on Plasma Physics

Series Editor: **Professor E W Laing**, University of Glasgow

*Other books in the series*

**An Introduction to Alfven Waves**
R Cross

**MHD and Microinstabilities in Confined Plasma**
W M Manheimer and C N Lashmore-Davies

**Transition Radiation and Transition Scattering**
V L Ginzburg and V N Tsytovich

*Forthcoming title*

**Plasma Diagnostics based on Forward Angle Scattering**
L Sharp, J Howard and R Nazikian

The Adam Hilger Series on Plasma Physics

# Radiofrequency Heating of Plasmas

R A Cairns

*Department of Mathematical and Computational Sciences,
University of St Andrews*

Adam Hilger
Bristol, Philadelphia and New York

*British Library Cataloguing in Publication Data*

Cairns, R. A.
  Radiofrequency heating of plasmas.
    1. Plasmas
    I. Title
    530.44

    ISBN 0-7503-0034-5

*Library of Congress Cataloging-in-Publication Data*

Cairns, R. A.
    Radiofrequency heating of plasmas/R. A. Cairns.
      p.  cm.—(The Adam Hilger series on plasma physics)
    Includes bibliographical references and index.
    ISBN 0-7503-0034-5 (hbk.)
    1. Plasma heating. 2. Tokamaks. I. Title. II. Series.
  QC718.5.H5C35  1991
  621.48′4—dc20                                 90-44386
                                                CIP

Published under the Adam Hilger imprint by IOP Publishing Ltd
Techno House, Redcliffe Way, Bristol BS1 6NX, England
335 East 45th Street, New York, NY 10017-3483, USA

US Editorial Office: 1411 Walnut Street, Philadelphia, PA 19102

Typeset by KEYTEC, Bridport, Dorset
Printed in Great Britain by J W Arrowsmith Ltd, Bristol

# Contents

**Preface**                                                                                    **vii**

**1  Introduction**                                                                              **1**

1.1   The use of radiofrequency heating in fusion research              1
1.2   Wave propagation in a plasma                                               3
1.3   Mode conversion                                                                      9
1.4   Particle diffusion                                                                      17
1.5   Transit time damping                                                                24

**2  Alfven Wave Heating**                                                                      **26**

2.1   Introduction                                                                              26
2.2   Properties of the Alfven wave and the heating mechanism      27
2.3   Effects of cylindrical and toroidal geometry                          33
2.4   Antenna–plasma coupling                                                          37
2.5   Experiments on Alfven wave heating                                       39

**3  Ion Cyclotron Heating**                                                                    **42**

3.1   Introduction                                                                              42
3.2   Wave propagation in the ion cyclotron range of frequencies   43
3.3   Heating at the second harmonic                                              46
3.4   The two-ion hybrid resonance and minority heating              52
3.5   Calculations in tokamak geometry and antenna coupling       55
3.6   Experiments on ion cyclotron heating                                     61
3.7   Ion Bernstein wave heating                                                      67

**4  Lower Hybrid Heating**                                                                     **71**

4.1   Introduction                                                                              71
4.2   Lower hybrid waves in a cold plasma and the
        accessibility condition                                                              72

4.3   Absorption of lower hybrid waves                               79
4.4   Launching of lower hybrid waves                                87
4.5   Experimental results on lower hybrid heating                   90
4.6   Fast wave heating                                              93

**5   Electron Cyclotron Heating                                     97**

5.1   Introduction                                                   97
5.2   Cold plasma propagation                                        98
5.3   Electron cyclotron wave propagation in a hot plasma           102
5.4   Non-linear effects at high powers                             111
5.5   Experimental aspects                                          115

**6   Current Drive                                                  120**

6.1   Introduction                                                  120
6.2   Basic principles                                              121
6.3   Langevin equations                                            124
6.4   Adjoint methods                                               127
6.5   Lower hybrid current drive                                    132
6.6   Electron cyclotron current drive                              138
6.7   Minority species current drive                                144
6.8   The anomalous Doppler instability                             146
6.9   Conclusion                                                    149

**References                                                         151**

**Index                                                              159**

# Preface

A reader who is at all acquainted with the large and rapidly growing literature on radiofrequency heating of plasmas will realise that a comprehensive survey of all the theory which has been developed, and the experiments which have been carried out, would require a very substantial volume. As a result, my aim in this book has been the less ambitious one of providing a readable introduction to the subject which describes the physical mechanisms underlying each of the methods used in radiofrequency heating, together with some of the main experimental results. I hope that the book will be intelligible to anyone who has made a study of the basic theory of plasma waves, and that after reading it they will be in a position to understand the research literature.

The plan of the book is that an introductory chapter acts as a reminder of the basic ideas of wave propagation in plasmas, with which an intending reader is assumed to be familiar, and describes some of the techniques and ideas which are common to all of the heating schemes. The following chapters give more detailed accounts of heating in the various frequency ranges, while the final chapter discusses radiofrequency current drive. The ideas are presented in the context of tokamak heating, since it is on such devices that most work has been done. Also, armed with a knowledge of radiofrequency heating in tokamaks, the reader should have no difficulty in understanding the application of similar ideas to stellarators or other devices.

Anything with any pretensions to be a complete bibliography of the subject would take up a large fraction of a book of this size, but I have tried to provide sufficient references for the interested reader to be able to gain entry to the literature. Omission of a paper does not mean that it is not to be regarded as an important contribution to the subject. Where there is a large literature on some topic I have tended to refer to some of the more recent papers, on the basis that they will lead the reader back to earlier work. Other omissions may, of course, simply be the result of ignorance, since I do not claim to be familiar with everything written on radiofrequency heating.

As is always the case with the author of such a work, I owe a great deal to colleagues, discussions with whom have played a large part in shaping my views on the subject. I shall not try to list everyone to whom I am indebted, but should make special mention of Chris Lashmore-Davies, who first persuaded me that this was an interesting area in which to work, and with whom I have enjoyed a fruitful collaboration over a number of years. I am also especially indebted to Vladimir Fuchs and Abraham Bers, whose hospitality I have enjoyed on a number of occasions and with whom I have had many interesting talks on different aspects of radiofrequency heating.

**R A Cairns**
April 1990

# 1 Introduction

## 1.1 The use of radiofrequency heating in fusion research

In a fusion reactor, the temperature required is of the order of 10–20 keV, this being the temperature range in which the cross section for the deuterium–tritium reaction is greatest. Radiofrequency heating is one of the main ways in which it is envisioned that such temperatures may be reached, and has already shown its potential in a great number of experiments. The discussion in this book will mostly deal with the tokamak, which is currently the most popular type of magnetic containment device and the likeliest candidate for development into a fusion reactor, but it should be borne in mind that radiofrequency heating and current drive can also be applied to other magnetic containment devices such as stellarators or mirrors. The magnetic field geometry in these is somewhat different, but the same basic principles apply.

A tokamak, details of which are described by Wesson (1987), is a toroidal device, and the main distinguishing feature of this particular type of machine is that the plasma carries a current in the toroidal direction, the standard mode of operation being for this current to be induced by a changing magnetic flux linking the torus. This current plays an essential role in the containment of the plasma, generating the poloidal field component necessary for the existence of a toroidal equilibrium. It also heats the plasma through ordinary resistive, or Ohmic, heating. A plasma, however, has an electrical resistivity which decreases with temperature, as $T^{-3/2}$, so as the temperature increases Ohmic heating becomes less effective, the value of the current being limited by requirements of stability of the plasma. For this reason, if the temperature is to be raised above around 3–4 keV, some form of auxiliary heating is needed.

There are two broad types of auxiliary heating which are used, namely neutral beam injection and radiofrequency heating. All large tokamaks use one, or in many cases both, of these. Neutral beam

1

heating involves the injection of a beam of high energy neutral atoms which can cross the confining magnetic field and then, through collisions and charge exchange reactions, transfer its energy to the plasma. Radiofrequency heating, which is the concern of this book, involves launching into the plasma high power electromagnetic waves, tuned to some natural resonant frequency of the plasma, which will lead to absorption of the wave and transfer of its energy to the plasma particles. For any viable scheme of radiofrequency heating the main requirements are that it should be possible to launch a wave from an antenna or waveguide at the plasma edge and that the wave can propagate to the central region of the plasma and be absorbed there.

As well as straightforward heating, radiofrequency beams may be useful in controlling the plasma profile. In Ohmic heating the current and energy deposition profiles are determined by the transport properties of the plasma and are not readily susceptible to external control. Radiofrequency absorption, on the other hand, may be localised and its position controllable, particularly in the higher frequency ranges, allowing it to be used to alter the temperature profile of the plasma in such a way as to tend to suppress instabilities. Another important application of radiofrequency waves is to current drive. The normal inductive method of driving the toroidal current in a tokamak leads, by its very nature, to pulsed operation since the driving field can only be maintained for as long as the magnetic flux linking the plasma torus is changing monotonically. In a reactor this would create very considerable engineering problems and for this reason much attention has been given to schemes to produce steady state current drive. One of the most important of these methods is the use of radiofrequency waves, which when sent into a tokamak at an angle to the major radius are capable of being absorbed in such a way as to set up a drift of electrons around the torus. Again, the fact that the position at which the waves are absorbed may be controlled can be exploited to change the current profile. Since some of the most important magnetohydrodynamic instabilities of a tokamak depend on this profile, this an an important application of wave current drive.

The schemes which have been investigated for radiofrequency heating of tokamaks fall into four main frequency ranges, within which waves excited at the plasma edge may propagate into the central region of the tokamak and be absorbed. The frequencies involved depend on the plasma parameters, and the values given in what follows are simply indications of what is typical in present-day machines. The lowest in frequency is the Alfven wave scheme, involing waves at a frequency of a few MHz. This is well below the ion cyclotron frequency of a typical tokamak, so the waves involved are the shear and compressional Alfven waves. Next comes the ion cyclotron range of frequencies (ICRF) at a few tens of MHz. At the moment this is the scheme which has been most widely used to heat large tokamaks, the JET (Joint European Torus) machine, for example, having some 16 MW of power available in this frequency range. Increasing the frequency to a few GHz brings us into the lower hybrid range, which has been particularly successful for current drive experiments. Finally we come to electron cyclotron heating at frequencies of around 30 GHz and upwards. Development of this method has, to some extent, been held back by the difficulty of manufacturing high power sources at the necessary high frequencies. However, in recent years there have been substantial advances in the development of gyrotrons, which generate the radiation required through a maser effect in which energy is extracted from a relativistic electron beam in a magnetic field and converted into electromagnetic energy. The other heating schemes are in frequency ranges where high power sources have already been developed for radar and other applications, so that the necessary technology already existed when these frequencies became of interest in plasma physics.

The remainder of this introductory chapter contains basic ideas on wave propagation and absorption which are common to all the frequency ranges. There then follow a series of chapters devoted to each of the frequency ranges listed above, describing the particular features of each and the main experimental results which have been obtained. In the final chapter the application of radiofrequency waves to current drive is discussed.

This book has been written as a short introduction to the subject and is intended to be accessible to non-specialists with only a general background in plasma physics. Further details can be found in the references given in the text and in the collections of review papers edited by Bers (1984) and Granatstein and Colestock (1985). The development of the subject can be followed in the proceedings of the series of Topical Conferences on Radio Frequency Power in Plasmas, published by the American Institute of Physics or of the International Symposia on Heating in Toroidal Plasmas, published by the Commission of the European Communities. A comprehensive review of current drive has been given by Fisch (1987).

## 1.2 Wave propagation in a plasma

This section is intended as a reminder of the basic properties of wave propagation in a hot plasma and as a convenient source of reference for the definition of quantities which we shall require subsequently. Derivations of most of the results given here can be found in the standard texts on plasma waves by Stix (1962) or Allis *et al* (1963) or in the more recent book by Swanson (1989). In a homogeneous plasma in a uniform magnetic field we follow the usual choice of coordinates in which the $z$ axis is taken along the direction of the field. With all wave quantities varying as $\exp(\mathrm{i}\boldsymbol{k}\cdot\boldsymbol{r} - \mathrm{i}\omega t)$, we obtain

$$\boldsymbol{n} \times (\boldsymbol{n} \times \boldsymbol{E}) = -\varepsilon\cdot\boldsymbol{E} \qquad (1.1)$$

where $\boldsymbol{n} = \boldsymbol{k}c/\omega$ is the plasma refractive index and $\varepsilon$ is the plasma dielectric tensor. The elements of the dielectric tensor are given by

$$\varepsilon_{ij} = \delta_{ij} + \sum_{s} \varepsilon_0 q_s^2/(m_s\omega^2) \sum_{n=-\infty}^{\infty} \int \mathrm{d}^3 v S_{ij}/(\omega - k_z v_z - n\Omega_s) \qquad (1.2)$$

the sum over $s$ being over the different particle species which may be present in the plasma, with charges $q_s$ and masses $m_s$. The quantity $\Omega_s$ is the cyclotron frequency of the species $s$, equal to $q_s B/m_s$. The $S_{ij}$ are given by

$$S_{ij} = \begin{pmatrix} v_\perp U(nJ_n/a_s)^2 & -\mathrm{i}v_\perp UnJ_nJ'_n & v_\perp WnJ_n^2/a_s \\ \mathrm{i}v_\perp UnJ_nJ'_n/a_s & v_\perp UJ_n'^2 & \mathrm{i}v_\perp WJ_nJ'_n \\ v_z UnJ_n^2 & -\mathrm{i}v_z UJ_nJ'_n & v_z WJ_n^2 \end{pmatrix} \qquad (1.3)$$

where

$$U = (\omega - k_z v_z)\, \partial f_{0s}/\partial v_\perp + k_z v_z\, \partial f_{0s}/\partial v_z$$

$$W = (n\Omega_s v_z/v_\perp)\, \partial f_{0s}/\partial v_\perp + (\omega - n\Omega_s)\, \partial f_{0s}/\partial v_z$$

$$a_s = k_\perp v_\perp/\Omega_s.$$

The arguments of the Bessel functions $J_n$ and their derivatives in (1.3) are $a_s$.

In (1.3), $f_{0s}$ is the velocity distribution function for the species $s$. Often it is reasonable to assume that the plasma is close to thermal equilibrium, or, at least, that the individual species have distribution functions close to Maxwellian, though possibly at different temperatures. In this case the velocity integrals in (1.2) can be evaluated and we obtain

$$\varepsilon_{ij} = \delta_{ij} + \sum_{s} (\omega_{ps}^2/\omega)(m_s/2T_s)^{1/2}\, \mathrm{e}^{-\lambda_s}/k_z \sum_{n=-\infty}^{\infty} T_{ij} \qquad (1.4)$$

with

$$T_{ij} = \begin{pmatrix} n^2 I_n Z/\lambda_s & -\mathrm{i}n(I'_n - I_n)Z & nI_n Z'/(2\lambda_s)^{1/2} \\ \mathrm{i}n(I'_n - I_n)Z & (n^2 I_n/\lambda_s + 2\lambda_s I_n - 2\lambda_s I'_n)Z & \lambda_s^{1/2}(I'_n - I_n)Z'/\sqrt{2} \\ nI_n Z'/(2\lambda_s)^{1/2} & \lambda_s^{1/2}(I'_n - I_n)Z'/2^{1/2} & I_n Z \end{pmatrix}$$

$$(1.5)$$

where

$$\lambda_s = T_s k_\perp^2 / (m_s \Omega_s^2)$$

with $T$ assumed to be in energy units (otherwise it should be multiplied by Boltzmann's constant). The modified Bessel function $I_n$ and its derivative have argument $\lambda_s$, and $Z$ is the plasma dispersion function with argument

$$\zeta_n = (\omega - n\Omega_s)(m_s/2T_s)/k_z.$$

Often the main features of wave propagation throughout much of the plasma can be obtained from the cold plasma approximation, in which the non-zero elements of the dielectric tensor are

$$\varepsilon_{11} = \varepsilon_{22} = 1 - \sum \omega_{ps}^2/(\omega^2 - \Omega_s^2) = \varepsilon_\perp$$

$$\varepsilon_{12} = -\varepsilon_{21} = -i\sum \omega_{ps}^2 \Omega_s/[\omega(\omega^2 - \Omega_s^2)] = i\varepsilon_{xy} \quad (1.6)$$

$$\varepsilon_{33} = 1 - \sum \omega_{ps}^2/\omega^2 = \varepsilon_\parallel.$$

To obtain this result from (1.5) we use the fact that $Z(x)$ behaves as $-1/x$ for large $x$, while for small $x$, $I_n(x)$ is approximately $x^n/2^n n!$. All terms except those with $n = 0$ or $\pm 1$ vanish as the temperature goes to zero. In all of these formulae it should be noted that the electron cyclotron frequency has to be taken to be negative, to preserve the symmetry of the equations. Hot plasma effects are usually important only in the vicinity of the region where the waves are absorbed. In such a region kinetic effects may, in addition to giving rise to damping, produce extra wave modes, like the Bernstein modes, which do not appear at all in cold plasma theory. In a homogeneous plasma the $x$ axis can be chosen, without any loss of generality, to lie along the perpendicular component of $n$, in which case the plasma dispersion relation becomes

$$\begin{vmatrix} \varepsilon_{11} - n_z^2 & \varepsilon_{12} & n_x n_z \\ \varepsilon_{21} & \varepsilon_{22} - n_x^2 - n_z^2 & 0 \\ n_x n_z & 0 & \varepsilon_{33} - n_x^2 \end{vmatrix} = 0. \quad (1.7)$$

If, however, there is an inhomogeneity in the plasma, then this may introduce a preferred direction in the plane perpendicular to the field. Often it is desirable to take one of the coordinate axes perpendicular to the field along the perpendicular density gradient. In this case it is obviously impossible in general to take the perpendicular component of $k$ along an axis and non-zero values of both $k_x$ and $k_y$ may need to be included in the local dispersion relation. The relevant dispersion relation may be obtained from (1.1), but will not be required in what we do.

Radiofrequency heating involves propagation in an inhomogeneous plasma, so it is necessary to consider the behaviour of waves in such a medium. Across the minor axis of a tokamak, in the equatorial plane, the plasma density has an approximately parabolic profile, while the toroidal magnetic field, which is the largest component, increases from the outside to the inside with a $1/R$ dependence, where $R$ is the major radius. Often the basic processes involved in radiofrequency heating of a tokamak can be studied using a simpler slab geometry, in which the system is taken to be inhomogeneous along only one direction, corresponding to the variation across the minor cross section. The plasma is assumed to be homogeneous in the other two directions, so the three-dimensional toroidal geometry of the machine is not taken into account in this approximation. For waves launched in the equatorial plane of the tokamak it is, however, a reasonable approximation which simplifies the analysis and enables us to understand many of the essential features of wave propagation in the more complicated system. If the wavelength of the radiation is short compared to the scale length of the tokamak, then the behaviour of the wave throughout much of the plasma may be well described by the Wentzel, Kramers, Brillouin (WKB) approximation, with the wave electric field taken to vary as

$$E_0(x)\exp\left(i\int^x k_x(x')\,dx' + ik_y y + ik_z z - i\omega t\right). \tag{1.8}$$

Here the $x$ axis is taken to be along the direction of the inhomogeneity. All other quantities associated with the wave of course behave in the same way. The wave propagates with fixed $k_y$, $k_z$ and $\omega$ and with $k_x(x)$ determined through the local dispersion relation, the parameters of which are functions of $x$. The variation of the field amplitude $E_0$ in (1.8) is given as follows (Berk and Book 1969). If the equation (1.1) is written in the form

$$K_{ij}E_j = 0$$

(with the usual convention that repeated tensor indices are to be summed over), then in the inhomogeneous system

$$\frac{\partial K_{ij}}{\partial k_x}\frac{dE_{0j}}{dx} + \frac{1}{2}\frac{d}{dx}\left|\frac{\partial K_{ij}}{\partial k_x}\right|E_{0j} = 0. \tag{1.9}$$

In a non-dissipative system $K_{ij}$ is Hermitian (i.e. equal to the complex conjugate of its transpose), so that if we take the product of (1.9) with $E_{0i}$ and add the result to its complex conjugate we get

$$\frac{d}{dx}\left(E_{0i}\frac{\partial K_{ij}}{\partial k_x}E_{0j}\right) = 0 \tag{1.10}$$

which is simply a statement of the conservation of the $x$ component of the wave energy flux (Stix 1962, Bers 1975).

In three dimensions, the extension of the WKB method is called the eikonal, or ray-tracing, approximation (Weinberg 1962, Bers 1975, Bernstein and Friedland 1983). The amplitude is assumed to vary as

$$E_0 \exp(i\Phi - i\omega t)$$

where $\Phi$ is a function of position such that $\nabla\Phi = k$, with $k$ satisfying the local dispersion relation. The direction of energy flow is given by the group velocity

$$v_g = \partial\omega/\partial k \tag{1.11}$$

that is, in Cartesian coordinates, $(\partial\omega/\partial k_x, \partial\omega/\partial k_y, \partial\omega/\partial k_z)$. If the background plasma in which the wave propagates is not time-dependent, then the frequency is constant and $k$ is determined by a dispersion relation of the form

$$D(\omega, k, r) = 0 \tag{1.12}$$

where $D$ is the determinant of $K_{ij}$. This single equation does not, however, determine the vector $k$ completely, and another equation is needed to determine how $k$ varies along the ray path. This can be derived from the fact that along the ray path

$$dD/dt = (\partial D/\partial k)(dk/dt) + (\partial D/\partial r)(dr/dt) = 0.$$

From (1.11)

$$dr/dt = -(\partial D/\partial k)/(\partial D/\partial \omega) \tag{1.13}$$

and so

$$dk/dt = (\partial D/\partial r)/(\partial D/\partial \omega). \tag{1.14}$$

Starting from a source at the plasma edge, producing a given wavenumber spectrum, equations (1.13) and (1.14) can be integrated to find the paths along which the various components of the spectrum flow into the plasma. The amplitude of the wave along the ray path is most easily found from the requirement of conservation of energy flux. Numerical codes incorporating this type of calculation are used extensively in the study of radiofrequency heating and current drive in tokamaks and a good review of the use of ray-tracing techniques is given by Brambilla (1986). The form of (1.13) and (1.14) is similar to Hamilton's equations in mechanics, and in systems with some degree of symmetry there may be conserved quantities, corresponding to the momenta conjugate to ignorable coordinates. In the present context these are coordinates on which $D$ does not depend, so that (1.14) means that the corresponding component of $k$ is conserved. Some applications of this arc given by Wersinger *et al* (1982). A tokamak in particular is axisymmetric, that is, its properties are independent of the position around the major axis. Thus, if $\phi$ is the toroidal angle, the waves may be written as a sum of components going as $\exp(in\phi)$, with $n$ taking integer values. These components propagate independently, since $n$ is a conserved quantity.

Returning to slab geometry and the WKB approximation, it is well known that there are various circumstances under which the approximation breaks down, even when the gradients are not too large. The most familiar are at a cut-off, where $k_x^2$ passes through zero, or a resonance, where $k_x^2$ goes to infinity. If these are isolated, then the wave is reflected at the cut-off and absorbed at the resonance (Stix 1962). In systems of practical interest it occurs quite commonly that there is both a cut-off and a resonance in the plasma, with a region between them where the wave is evanescent (i.e. $k_x^2 < 0$). If the distance between the cut-off and the resonance is not too large, it is possible for part of the wave energy to tunnel through the evanescent region. This was considered by Budden (1961) who pointed out that, since the back-to-back resonance and cut-off could be described most simply by a wavenumber dependence of the form

$$k^2 = a + b/x$$

so that the resonance is at $x = 0$ and the cut-off at $x = -b/a$, then the wave amplitude in this region may be expected to be given by an equation of the form

$$\mathrm{d}^2\phi/\mathrm{d}x^2 + (a + b/x)\phi = 0. \tag{1.15}$$

An equation of this form, which in the present context has become known as Budden's equation, is a special case of a standard equation whose solution is expressible in terms of confluent hypergeometric functions. The asymptotic properties of the solution, which tell us how the WKB solution representing an outgoing wave from one side is connected to a solution representing a superposition of incoming and reflected waves on the other side, are readily available and can be used to obtain the transmission and reflection coefficients and the fraction of the incident energy absorbed at the resonance. The result of such a calculation is that with incidence from either side the fraction of the energy which is transmitted is

$$T = \exp(-2\pi|b|). \tag{1.16}$$

If the wave is incident on the cut-off a fraction

$$R = (1 - T)^2 \tag{1.17}$$

of the energy is reflected, while if it is incident on the resonance there is no reflection. In either case that which is not transmitted or reflected is absorbed at the resonance. Although no absorption mechanism is included in (1.15), a small amount has to be added to resolve the singularity in this equation. The amount of damping is independent of the damping coefficient, the function of which is simply to move the singularity slightly off the real axis and so determine the way in which the solution is continued around it

The WKB approximation also breaks down if in some region of the plasma there are two values of the wavenumber which are close together. At such a region two waves, which elsewhere in the plasma are well defined separate modes, propagating independently, come together and can interact with each other. The result is mode conversion, where a wave incident on this region from one side emerges as a combination of the two modes. This process has attracted a great deal of attention in the context of radiofrequency heating, since many of the heating schemes involve a process of this sort, with one mode excited by an antenna at the plasma edge and being converted in the interior of the plasma to another mode which is, in turn, absorbed. It should be noted that this is a purely linear phenomenon. We are not talking about a non-linear interaction between modes at different frequencies, but about a single solution of the linear wave problem at a single frequency. The problems arise simply because the solution cannot be represented everywhere by a single WKB type solution. Because of the importance of mode conversion we shall devote a separate section to a discussion of some of the techniques which have been developed for its analysis.

## 1.3 Mode conversion

Mode conversion is a linear process occurring in an inhomogeneous plasma in the vicinity of a region where two solutions of the dispersion relation for a given frequency have almost the same wavelength. Considering a purely one-dimensional system, for convenience, the behaviour of the wavenumber as a function of position is typically as shown schematically in figure 1.1. The background plasma is assumed to be time-independent, so that waves propagate with a fixed frequency. In the slab geometry which we have already described as a reasonable first approximation to a tokamak, the wavenumber components perpendicular to the direction along which the density and magnetic field vary are constant. In essence this is a one-dimensional wave propagation problem, the important wavenumber component being that parallel to the gradients. The other components can be regarded as constant parameters which do not need to be considered explicitly at the moment.

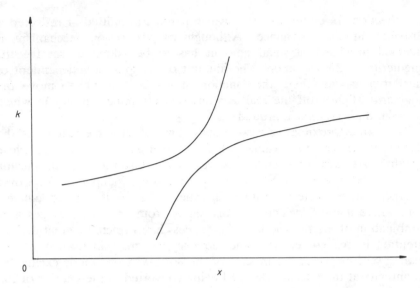

**Figure 1.1** The dependence of wavenumber on position in a region where mode conversion occurs.

Away from $x_0$, the position at which the two roots of the local dispersion relation are closest, there is a straightforward superposition of two wave modes, each of which can be described by the WKB approximation. However, around $x_0$ the two modes are no longer distinct. If energy is incident on this region in only one of the modes, the outgoing flow of energy will be in a superposition of the two modes. In radiofrequency heating the importance of this process is that the absorption of energy often does not take place directly from the mode excited by the source, but instead takes place after it has been converted into the other mode. The basic problem is to find the fraction of the energy which is transmitted in the original mode and the fraction which is mode converted.

The obvious way to do this is to abandon the WKB approximation and look for a solution of the exact equations describing the wave propagation in the inhomogeneous plasma. Away from the mode conversion region these would be expected to have an asymptotic form which could be identified with the WKB solutions. However it has to be remembered that in most cases the absorption process involves modes described by the warm plasma dispersion relation and that a solution of the linearised Vlasov–Maxwell system of equations in an inhomogeneous plasma is very difficult to obtain. For this reason various approximation techniques have been devised, with a view to describing the wave propagation by a differential equation rather than an integro-differential equation.

One way of doing this is to simplify the local dispersion relation, generally by making a small Larmor radius expansion, so that it becomes a polynomial in the wavenumber. The argument of the Bessel functions in the dispersion relation is the Larmor radius times the perpendicular wavenumber, so this involves taking the first few terms in the series expansion of these functions. A differential equation may then be constructed through the so-called inverse Fourier transform

$$k \rightarrow -\mathrm{i}d/dx.$$

Bearing in mind that waves propagate in both directions, it can be seen that to produce a diagram like figure 1.1 there will be at least two other roots for $k$, negative values corresponding to propagation in the opposite direction, so that the dispersion relation will be of at least fourth order in $k$. A straighforward inverse Fourier transform then yields an equation of fourth or higher order. For instance to describe mode conversion from a fast magnetosonic to a Bernstein mode, Ngan and Swanson (1977) proposed the equation

$$y^{\mathrm{iv}} + \lambda^2 x y'' + (\lambda^2 x + \gamma)y = 0 \qquad (1.18)$$

with $\lambda$ and $\gamma$ constants. For large $x$ the wavenumbers are approximately $k^2 = 1$ and $k^2 = x$, the first of which represents the fast mode whose wavenumber is approximately constant in the region of interest and can, with suitable scaling of the length, be taken to be one. The other root is evanescent for $x < 0$ (assuming $\lambda > 0$) and for positive $x$ gives a WKB solution going as

$$\exp\left(\tfrac{2}{3}\, \mathrm{i}\lambda^{1/2}x^{3/2}\right).$$

This approximates the behaviour of the Bernstein mode which has a cut-off at the cyclotron harmonic.

Equations like (1.18), where the coefficients are linear in $x$, may be solved using the method of Laplace, in which a solution of the form

$$y(x) = \int_C F(k) \exp(-\mathrm{i}kx)\, dx \qquad (1.19)$$

is sought. If the end points of the contour C are chosen so that the integrand vanishes there, then integration by parts shows that

$$xy(x) = \int_C \mathrm{i}F'(k) \exp(-\mathrm{i}kx)\, dx,$$

while derivatives of $y$ involve multiples of the transform by powers of $\mathrm{i}k$ in the usual way. Substituting into the equation shows that it is satisfied if a first-order equation in $F(k)$ is satisfied. The contours are usually chosen so that their end points go off to infinity along directions in which the integrand tends to zero. Generally there are a number of sectors in the complex plane satisfying this condition, and the different linearly independent solutions of the equation can be constructed by choosing the end points in different ways. The solution of interest is determined by the fact that its asymptotic properties must represent the combinations of ingoing and outgoing waves appropriate to the problem. The form of the solution (1.19) is such that its asymptotic behaviour can be found by the method of steepest descents. The saddle points in the complex plane are found in the standard way and the contribution to the asymptotic solution from a saddle point gives one of the WKB solutions. The asymptotic solutions correspond to superpositions of the WKB solutions for the separate waves, but by this method a complete solution is obtained which gives the connection through the region in which the WKB solution is not valid. The contour is chosen so as to pass through the saddle points which give the required behaviour. A general formulation of this method, applicable to $n$th-order equations with linear coefficients, was given by Gambier and Schmitt (1983), and further discussion of how to choose the appropriate path of integration and do the asymptotics is given in a review article by Swanson (1985).

A problem with generating the differential equation in the simple way described above is that the inverse Fourier transform is not unique. To see this we may simply note that the terms

$$a\,\mathrm{d}y/\mathrm{d}x \qquad \text{and} \qquad \mathrm{d}(ay)/\mathrm{d}x$$

give the same Fourier transform if $a$ is a coefficient which is constant in a homogeneous system. If however we go to an inhomogeneous system, where $a$ is a function of position, then they are quite different. From the homogeneous plasma dispersion relation it is impossible to tell which is correct. Later work by Swanson (1981) and others (e.g. Colestock and Kashuba 1983) has avoided this problem by carrying out a more systematic derivation of approximate differential equations to describe the behaviour of waves in the mode conversion region. This involves starting from the Vlasov–Maxwell set of equations and making a small Larmor radius expansion to simplify them in a systematic way, obtaining a differential equation in place of the original integro-differential system. Some more details of how this is done are given in Chapter 3, in the context of a more specific problem of ion cyclotron heating.

An alternative line of approach is to recognise that in a mode conversion process of the type represented by figure 1.1 only two waves are interacting, and that the backward propagating waves play no role. It might then be thought that a second-order equation could be constructed to give the required information. Such ideas have quite a long history, a well known example being the work of Heading (1961) in which it was shown how a higher-order system of differential equations can be transformed in such a way that in the vicinity of a mode conversion point a second-order system separates out and is only weakly coupled to the remaining equations. Away from the mode conversion point the second-order equation has solutions representing a superposition of the two waves involved in the mode conversion. Since a hot plasma is described by a complicated integro-differential system, work in this area has largely concentrated not on reducing the exact equations, but on obtaining suitable equations from the local dispersion relation. Theories of this sort were proposed by Fuchs *et al* (1981, 1985a) and by Cairns and Lashmore-Davies (1982, 1983). A comparison of these two theories, which are similar in some respects, is given by Lashmore-Davies *et al* (1985). We shall give here a description of the version due to Cairns and Lashmore-Davies, if only because it is more familiar to the author.

The essential feature of this method is the fact that in the vicinity of a point where the dispersion curves cross over, as in figure 1.2, the dispersion relation can be approximated by

$$(k - k_1)(k - k_2) = \mu \tag{1.20}$$

where $k = k_1(x)$ and $k = k_2(x)$ give the wavenumbers of the uncoupled modes, as shown in figure 1.2, while $\mu$ is a measure of the closest distance between the actual curves. The aim is now to associate with this local dispersion relation a differential equation which will, it is hoped, give an accurate description of the behaviour around $x_0$. Clearly we wish

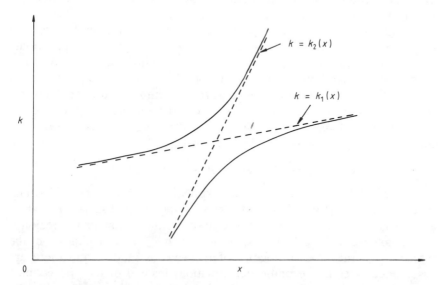

**Figure 1.2** Wavenumber dependence of the uncoupled modes used in the method of Cairns and Lashmore-Davies.

the differential system to reproduce the approximation (1.20) to the dispersion curves, but, as has already been pointed out, this condition is not sufficient to give a unique result.

Cairns and Lashmore-Davies suggest that in a non-dissipative system, the way to choose the correct differential equation is to look for one which has a conservation law compatible with the energy conservation law of the system. For example if all the group velocities of the waves shown in figure 1.2 are in the positive direction, then if $\phi_1$ and $\phi_2$ are the amplitudes of the two modes, normalised so that $|\phi_1|^2$ and $|\phi_2|^2$ give the energy fluxes in the two modes, an obvious energy conservation law is

$$|\phi_1|^2 + |\phi_2|^2 = \text{constant.} \tag{1.21}$$

Here $\phi_1$ and $\phi_2$ cannot be identified simply as field amplitudes, as assumed by Swanson (1985) in a discussion of this method, but will also involve the dielectric tensor elements in such a way as to produce, in the WKB regime, the expression for the energy flux given by Stix (1962) and quoted above.

A differential system satisfying (1.21) and reproducing the dispersion relation (1.20) is

$$d\phi_1/dx - ik_1\phi_1 = i\mu^{1/2}\phi_2$$
$$d\phi_2/dx - ik_2\phi_2 = i\mu^{1/2}\phi_1. \tag{1.22}$$

With the further approximation that $k_1$ and $k_2$ are well approximated by linear functions of $x$ in the region of interest, one of the variables in the pair (1.22) can be eliminated, and the resulting equation transformed into Weber's equation, the solution of which, with the appropriate asymptotic behaviour representing one incoming and two outgoing modes, is the parabolic cylinder function. The asymptotic properties of this function are readily available and from them explicit expressions for the transmission and mode conversion coefficients can be obtained. If, in the region of mode conversion, $k_1 = a(x - x_0)$ and $k_2 = b(x - x_0)$, then the energy transmission coefficient is given by

$$T = \exp(-2\pi\mu/|a - b|).$$

This method thus provides a simple expression for the transmission and mode conversion coefficients, in terms of the local dispersion relation and the gradients in the neighbourhood of the mode conversion region.

The precise form of the equations to be taken depends on the directions of the group velocities of the various modes, and there are cases where a single second-order equation, instead of the two coupled equations of (1.22), gives the natural description. This possibility was recognised by Fuchs *et al* (1985b) and an explicit example was analysed in detail by Woods *et al* (1986). This last relates to lower hybrid heating and is described in more detail in Chapter 4.

In most practical applications, one of the waves has a cut-off near the mode conversion point, which means that the energy going into this mode can be reflected and couple to the oppositely directed waves. It appears, however, that even if the mode conversion region and the cut-off are not well separated in space, the behaviour of the whole system can be found by treating each mode conversion and cut-off separately and tracing the flow of energy along the various branches. Thus with dispersion curves as shown in figure 1.3, it was pointed out by Cairns and Lashmore-Davies (1983) that the results of fourth-order calculations, which automatically include the waves propagating in both directions, were reproduced by the simple argument illustrated in this figure, which predicts that the reflection and transmission coefficients are related by

$$R = (1 - T)^2$$

in a system with no damping. This result makes use of the fact that the transmission coefficient is the same, regardless of which mode is incident on the conversion region. The same kind of argument can be applied no matter in which wave the energy is incident, and a more complicated example which can be seen to be in accord with this is considered by Mjølhus (1987). A really convincing mathematical demonstration that

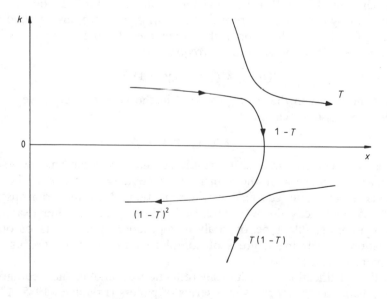

**Figure 1.3** Energy flow through a mode conversion and cut-off.

this sort of procedure is valid has not been produced, but it appears to work correctly in every case where it has been compared with solutions of higher-order equations. The main limitation is that the system has to be non-dissipative.

As an example of the use of this method we shall consider the equation of Ngan and Swanson (1977) which is quoted above. The dispersion relation corresponding to this equation is

$$k^4 - \lambda^2 x k^2 + \lambda^2 x + \gamma = 0$$

and to apply the theory we note that for large $x$ the roots are

$$k^2 \approx \lambda^2 x - 1 \quad \text{and} \quad k^2 \approx 1.$$

These roots give the behaviour of the solutions away from the mode conversion zone, and so correspond to the dotted lines in figure 1.2. Following the procedure outlined above, we write the dispersion relation as

$$(k^2 - 1)(k^2 - \lambda^2 x + 1) = -(\gamma + 1).$$

The crossing point of the uncoupled modes is then given by

$$\lambda^2 x - 1 = 1$$

i.e.

$$x = 2/\lambda^2$$

at which point $k^2 = 1$. We now expand about the coupling point $k = 1$, noting that by choosing this root rather than $k = -1$ we have picked out waves going in one direction rather than the other. Letting $\delta = k - 1$ and $\zeta = x - 2/\lambda^2$ we obtain the approximation

$$\delta(\delta - \tfrac{1}{2}\lambda^2 \zeta) = -(\gamma + 1)/4.$$

Identifying this with the form proposed in the general theory we arrive at the transmission coefficient

$$T = \exp\left[\pi(\gamma + 1)/\lambda^2\right].$$

This is precisely the same as the result obtained by Ngan and Swanson, by means of analysis of the fourth-order equation. This analysis also predicts a reflection coefficient which is related to the transmission coefficient in exactly the way described above. Various other examples in which this simple type of analysis reproduces the results of other calculations of varying degrees of complexity are given by Cairns and Lashmore-Davies (1983).

Further justification of such order reduction techniques has been given by Friedland and co-workers in a series of papers (Freidland 1985, 1986, Friedland and Goldner 1986, Friedland *et al* 1987, Friedland and

Kaufman 1987) and also by Cairns and Lashmore-Davies (1986a). In the first group of papers mentioned here a technique which the authors refer to as 'congruent reduction' is developed. The interested reader is referred to these papers for details, but a brief outline of the ideas involved is as follows. The theory starts from the Vlasov–Maxwell system of equations and shows how, with the assumption of slowly varying parameters and the neglect of damping, it is usually possible to make a transformation which reduces the system to a single first-order equation describing the variation of a suitable amplitude along a ray path. This is just equivalent to the standard eikonal approximation, and the equation which is obtained can be shown to be an expression of the conservation of energy flux along the ray. However, in the mode conversion region the system becomes degenerate and the transformation which singles out individual wavemodes becomes singular. It is shown how to modify the transformation in such a way that the two modes which are degenerate no longer separate out completely, but are described by equations of the type proposed on more heuristic grounds by Cairns and Lashmore-Davies or by Fuchs and co-workers. The result is a systematic technique which reduces the order of the equations describing a mode conversion problem by separating out those amplitudes which are coupled. It also gives an unambiguous interpretation of these amplitudes in terms of the electric field in the plasma, and has been extended to deal with more complicated geometries than the earlier theories, which confined themselves to a simple slab geometry. An extension to include collisionless damping described by the Vlasov equation has been given by Ye and Kaufman (1988) and involves the introduction of singular normal modes as in Van Kampen–Case theory.

Although the work referred to here deals mainly with non-dissipative plasmas, in practice it is usually the case that at least one of the waves involved is strongly damped, indeed the whole idea is to have damping of the wave energy in order to heat the plasma. A good deal of work has been devoted to this case, both to the solution of fourth- and higher-order systems of equations, and to the development of approximate methods to reduce the order of the equations. Since most of this has been directed towards ion cyclotron heating, it will be more convenient to defer discussion to the chapter on this topic, and discuss the methods within this particular context.

## 1.4 Particle diffusion

Having considered wave propagation we now consider the effect that the waves have on the particles. The particle velocity distribution in the course of radiofrequency heating is determined by the balance between the effect of the waves and the effect of collisions and of the DC electric field which will generally be present in a tokamak. The effect of the waves is generally to interact with some resonant group of particles, diffusing them in velocity space so as to flatten out the particle distribution function in the region of the resonant particles. The collisions, on the other hand, tend to restore the equilibrium Maxwellian distribution. The effect of collisions may be described by the usual Fokker–Planck collision term (Landau 1936, Rosenbluth *et al* 1957), though if the waves are resonant with high velocity particles it may be possible to use a simplified high velocity limit (Fisch 1987).

The velocity distribution functions of the different particle species within the plasma will be determined by the balance between the effect of the waves and the effect of collisions. To calculate each distribution function we require an equation of the form

$$\partial f/\partial t = (\partial f/\partial t)_w + (\partial f/\partial t)_c \qquad (1.23)$$

where $(\partial f/\partial t)_c$ is the collision term and $(\partial f/\partial t)_w$ is a term due to the waves. We shall discuss some of the properties of this equation later, particularly in the chapter on current drive, while here we confine our attention to the nature of the wave term.

An important point to note is that during wave absorption it is not generally the case that particles gain energy from the waves all the time. Whether a specific particle gains or loses energy depends on the phase relation between its position on its orbit and the wave, and either is equally likely. At different points in its orbit a particle may gain or lose energy, so the effect is not of a steady gain in energy by the particle, but rather of a random walk in velocity space. So far as the distribution of particles in velocity is concerned, such a random walk is equivalent to diffusion in velocity space which, as with any diffusion process, has the effect of levelling out the particle distribution and tending to produce a plateau in the region of velocity space on which the waves act.

A steady state solution of (1.23) is impossible if there is non-zero absorption of energy from the waves, since the correct form of the collision term conserves energy. However, it is usually possible to make use of a steady state approximation by recognising that the shape of the resonance region in velocity space reacts very quickly to the waves, and takes up a form in which the input of energy from the waves is balanced by the loss of energy to the bulk of the particles. The energy fed into the system from the waves produces, in the absence of any loss process, a steady increase in the bulk temperature. However, this takes place on a much longer timescale, a fact which allows useful results to be obtained with an approximation in which the Rosenbluth potentials are calculated using an assumed Maxwellian distribution with a given temperature. The steady state distribution and resulting wave damping can then be calculated from (1.23). In effect we are calculating the form of the distribution function in the resonant region on the assumption that it results from the balance between energy fed into the particles by the waves and energy lost to the bulk of the distribution which is at a given temperature. On a longer timescale the energy input can be fed into transport equations and used to update the bulk temperature.

Since the effect of waves on the particles is to produce diffusion, we can expect the term in the equation giving the rate of change of the distribution function resulting from the waves to take the form

$$(\partial f/\partial t)_w = \frac{\partial}{\partial \boldsymbol{v}} \cdot \mathbf{D} \cdot \partial f/\partial \boldsymbol{v} \qquad (1.24)$$

where $\mathbf{D}$ is a tensor proportional to the square of the wave amplitude. This is precisely the type of term obtained in the well known quasi-linear theory, where the form of the diffusion coefficient in a magnet-ised plasma has been given by Kennel and Engelmann (1966).

Often work on radiofrequency heating has simply used this form, with the geometrical effects of the toroidal shape of the tokamak introduced in an approximate way, for example by multiplying the diffusion coefficient by the ratio of the area of the part of a flux surface on which absorption takes place to the total area. However, work has been done more recently on deriving diffusion terms which take account of non-uniform fields (Bernstein and Baxter 1981) and the effect of localised irradiance of a tokamak (Demeio and Engelmann 1986, Cairns and Lashmore-Davies 1986b, O'Brien *et al* 1986). Probably the most sophisticated work on these lines is that of Kerbel and McCoy (1985) who have developed an elaborate computational code for the calculation of the diffusion coefficient averaged over a flux surface in a tokamak.

Before giving any further consideration to the diffusion coefficient, it is useful to consider the motion of a single particle in a wave. An understanding of this is important in understanding the absorption mechanism of the waves and how the distribution function is affected. Thus we consider the equation

$$\mathrm{d}\boldsymbol{v}/\mathrm{d}t = q/m(\boldsymbol{v} \times \boldsymbol{B}_0 + \boldsymbol{E} + \boldsymbol{v} \times \boldsymbol{B}) \qquad (1.25)$$

where $\boldsymbol{B}_0$ is the background magnetic field, assumed to be steady and uniform, and $\boldsymbol{E}$ and $\boldsymbol{B}$ are the wave fields, varying as $\exp(\mathrm{i}\boldsymbol{k}.\boldsymbol{r} - \mathrm{i}\omega t)$. We choose axes so that $\boldsymbol{B}_0$ is along the $z$ axis and $\boldsymbol{k} = (k_\perp, 0, k_z)$ is in the $x$–$z$ plane. The wave magnetic field can be eliminated through the relation

$$\boldsymbol{k} \times \boldsymbol{E} = \omega \boldsymbol{B}$$

and the equations simplified by decomposing them into right and left circularly polarised components, rather than Cartesian components, in the $x$–$y$ plane. Thus we define

$$v_{\pm} = v_x \pm iv_y$$

$$E_{\pm} = E_x \pm iE_y$$

and obtain from (1.25)

$$d/dt(v_+e^{i\Omega t}) = (q/m)e^{i\Omega t}[(1 - k_z v_z/\omega)E_+$$
$$- \tfrac{1}{2}(k_\perp v_+/\omega)(E_+ - E_-) + k_\perp v_z E_z/\omega]$$
$$\tag{1.26}$$

$$d/dt(v_-e^{-i\Omega t}) = (q/m)e^{-i\Omega t}[(1 - k_z v_z/\omega)E_-$$
$$+ \tfrac{1}{2}(k_\perp v_-/\omega)(E_+ - E_-) + k_\perp v_z E_z/\omega]$$
$$\tag{1.27}$$

and

$$dv_z/dt = (q/m)[(1 - k_\perp v_x/\omega)E_z + \tfrac{1}{2}(k_z/\omega)(E_+v_- + E_-v_+)] \tag{1.28}$$

where $\Omega = qB_0/m$ is the cyclotron frequency.

On the right-hand side of (1.26)–(1.28) we now use the unperturbed orbits

$$v_{\pm} = v_\perp e^{\mp i\Omega t} \tag{1.29}$$

and

$$E = E_0 \exp\{i[(k_\perp v_\perp/\Omega)\sin\Omega t + k_z v_z t - \omega t]\}$$

$$= E_0 \sum_n J_n(k_\perp v_\perp/\Omega)\exp(in\Omega t + ik_z v_z t - i\omega t) \tag{1.30}$$

the initial conditions having been chosen to eliminate the constants of motion which would appear in the general case.

Substituting (1.29) and (1.30) into (1.26) we obtain

$$d/dt(v_+e^{i\Omega t}) = (q/m)e^{i\Omega t}[(1 - k_z v_z/\omega)E_+ - \tfrac{1}{2}k_\perp v_\perp e^{-i\Omega t}(E_+ - E_-)/\omega$$

$$+ (k_\perp v_z/\omega)E_z]\sum_n J_n(k_\perp v_\perp/\Omega)\exp(in\Omega t + ik_z v_z t - i\omega t)$$

$$\tag{1.31}$$

where the components of $E$ represent the field amplitude, the subscript 0 having been dropped to avoid complicating the equation still further. A secular change in $v_+$ (i.e. a change which increases linearly with time) will occur if the right-hand side of (1.31) contains a constant term. This will happen if

$$\omega - n\Omega - k_z v_z = 0 \tag{1.32}$$

for some integer $n$, the constant part of the right-hand side of (1.31) being proportional to

$$(1 - k_z v_z/\omega)E_+ J_{n-1} - (\tfrac{1}{2}k_\perp v_\perp/\omega)J_n(E_+ - E_-) + (k_\perp v_z/\omega)E_z J_{n-1} \tag{1.33}$$

with the argument of the Bessel functions as before. Particles satisfying (1.32) are in resonance with the wave. Their response to the wave is not an oscillating velocity perturbation, but a steadily increasing change in velocity. This means that they either take energy from the wave or feed energy into it, depending on the sign of the change in velocity.

The wave and plasma parameters for radiofrequency heating are generally such that if $v_{th}$ is the thermal velocity of the species of interest, then

$$k_z v_{th}/\Omega \ll 1.$$

This means that for typical particle velocities the resonance condition (1.32) is satisfied for

$$\omega \approx n\Omega.$$

It is also often the case that $k_\perp v_{th}/\Omega \ll 1$ and the argument of the Bessel functions is much less than unity. There are some important problems, for example the heating of ions by lower hybrid waves (see Chapter 4), where this is not the case, but we shall not concern ourselves with them at the moment. Assuming that the inequality is satisfied, the value of the Bessel functions decreases rapidly as their order increases $(J_n(x) \approx x^n/(2^n n!)$ for small $x$). For ions $\Omega > 0$ and $n$ is positive, so, if we exclude for the moment the case $n = 0$, the dominant terms in (1.33) are those involving $J_{n-1}$ which are proportional to

$$E_+ + (k_\perp v_z/\omega)E_z.$$

The driving term for $v_-$ involves terms containing $J_n$ and $J_{n+1}$ and so is smaller, as is that for $v_z$.

Thus for ion cyclotron heating, the effective driving field is $E_+ + k_\perp v_z E_z/\omega$ which drives up the velocity component $v_+$. The physical reason for this is that $E_+$ is the left circularly polarised component of the field and rotates in the same direction as the basic ion cyclotron orbit. The particle then sees a constant field in its rest frame and the result is that there is a resonant interaction, with the particle velocity changing steadily with time. The origin of the second part of the driving term is not so obvious, involving as it does a longitudinal electric field, which drives the perpendicular velocity component. In fact this part of the driving force arises from the $v \times B$ force on the particle, as is evident from its velocity dependence. Since the ion cyclotron frequency in any tokamak is extremely small compared to the electron plasma frequency, electrons can flow along the magnetic field lines to cancel the parallel electric field. The result is that in ion cyclotron heating the term involving the parallel electric field does not enter. It may, however be important in electron cyclotron resonance heating.

For electron cyclotron resonance, the change in sign of the charge makes $v_-$ the most strongly driven velocity component, and the dominant driving term is

$$E_- + k_\perp v_z E_z / \omega.$$

Although the $E_z$ part of this is multiplied by a small quantity, it is necessary to include it since it is the dominant component for a wave in the ordinary mode.

In the case $n = 0$, i.e. $\omega = k_z v_z$, we have the familiar Landau damping condition and the acceleration is predominantly along the parallel direction. As we shall see later, this is the process which is of most importance in the lower hybrid frequency range.

To see how this kind of interaction between a wave and a particle gives rise to diffusion of a particle in velocity space, note first that we have made a particular choice of the phase of the wave, and of the particle in its cyclotron orbit. If we consider, for the sake of argument, a cyclotron resonance, so that it is the perpendicular degree of freedom which is of interest, then the behaviour of the particle will be given by

$$dv_+/dt = (q/m)E_{\text{eff}} J_{n-1} \cos \Phi$$

where we have assumed a positive ion, $E_{\text{eff}}$ is the effective driving field discussed above and $\Phi$ is an angle which depends on the relation between the phase of the wave and that of the particle in its orbit.

If we suppose that the cyclotron wave is excited by a number of antennae around the tokamak, producing a number of localised regions in which the particle is excited as it moves around the tokamak, then if $\tau$ is the time it takes the particle to pass through the wave beam, assumed uniform over its cross section for simplicity, the change in $v_+$ is

$$\Delta v_+ = (q/m)E_{\text{eff}} J_{n-1} \tau \cos \Phi.$$

Assuming that the phase relation between the particle and the wave is uncorrelated on successive transits of the beam, then after $N$ transits we have

$$\langle \Delta v_+^2 \rangle = \tfrac{1}{2} N |(q/m)E_{\text{eff}} J_{n-1} \tau|^2.$$

The number of beam transits will be proportional to time, so we have a characteristic random walk in which the mean square deviation of the particle from its initial velocity is proportional to time, while the mean deviation is zero. If we consider an ensemble of particles, described by a velocity distribution function, then the random walk of an individual particle in velocity space translates into a diffusion of the density of particles in velocity space (see for example Chandrasekhar 1943). The diffusion coefficient is one half of the rate of increase of the mean square deviation.

In calculating the diffusion coefficient some care is required since in the theory we deal with complex quantities representing the circularly polarised components, while the quantity with which we really need to deal is $v_\perp$, the physical value of the perpendicular velocity component. The relation between $\langle |v_\perp|^2 \rangle$ and $\langle |v_+|^2 \rangle$ can be seen as follows. We first note that $v_-$ is approximately constant, so that

$$\delta v_x = i \delta v_y$$

and

$$(\delta v_\perp)^2 = (\delta v_x)^2 + (\delta v_y)^2 = 2(\delta v_x)^2 = \tfrac{1}{2}|\delta v_+|^2.$$

A more detailed calculation is, of course, necessary to take into account the variation of the wave amplitude across the beams, the variation of the magnetic field along the orbit, the toroidal geometry of the system, including the effects of trapped particles, and so on. However, the above gives a general idea of how the behaviour of an individual particle can be related to the diffusion coefficient of a particle distribution. More detailed calculations along these lines can be found in Cairns and Lashmore-Davies (1986b), O'Brien *et al* (1986) and Demeio and Engelmann (1986).

We have concluded in the above that in a cyclotron resonance the particle is accelerated mainly in the perpendicular direction, while in a Landau resonance it is accelerated in the parallel direction. The same conclusions can also be obtained in quite a simple and instructive way from a consideration of momentum conservation. Since particles are free to move along the magnetic field lines, momentum along this direction must be conserved. For the wave,

$$\text{momentum/energy} = k/\omega$$

and so if the wave interacts with a particle, the changes in energy and parallel momentum of the particle must satisfy

$$\Delta p_z / \Delta E = k_z / \omega. \tag{1.34}$$

For a Landau resonance, with $\omega = k_z v_z$, (1.34) implies that

$$\Delta E = v_z \Delta p_z$$

which is only possible if the change in velocity is in the parallel direction alone. For a cyclotron resonance

$$\omega = n\Omega + k_z v_z$$

and so

$$\Delta p_z / \Delta E = k_z / (n\Omega + k_z v_z).$$

Using the fact that

$$\Delta E = m v_\perp \Delta v_\perp + m v_z \Delta v_z$$

this gives

$$v_\perp \Delta v_\perp = n\Omega \Delta v_z / k_z \qquad (1.35)$$

a geometrical interpretation of which is shown in figure 1.4. The change in velocity of a particle is along the circle in velocity space centred at the parallel phase velocity of the wave, and for $k_z v_z \ll \omega$ it is very close to the perpendicular direction, in agreement with our earlier conclusions. For an ensemble of particles interacting with the waves, particles may go in either direction along the allowed paths, depending on the phase relation between the particle gyro orbit and the wave. The paths are thus to be regarded as the directions in velocity space along which diffusion takes place.

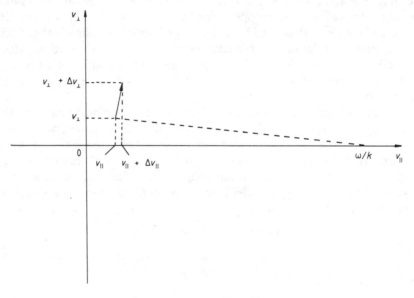

**Figure 1.4** Geometry of the direction of particle diffusion due to cyclotron resonance.

## 1.5 Transit time damping

This effect was first proposed as a heating scheme at low frequency, below the ion cyclotron range, using a periodic modulation of the confining magnetic field over a part of the tokamak. The name comes from the fact that, when the collision frequency is small, the optimum power input is obtained when the wave frequency is around the reciprocal of the average ion transit time through the region in which the field is modulated (Berger *et al* 1958). This did not prove to be a good way of heating ions, but the name survives and is used to refer to a damping mechanism which acts on waves, but involves the same physical effect as heating by a modulated confining field. It is now more often encountered in the context of electron heating by ion cyclotron waves, which are at a frequency well below the electron cyclotron frequency. The essential idea is based on the well known guiding centre theory of particle orbits, in which there is a force along the magnetic field lines towards regions of weaker field, given by

$$F = -\frac{\mathrm{d}}{\mathrm{d}z}(\mu B)$$

where $\mu$ is the conserved magnetic moment of the particle. If there is a wave involving a periodic modulation of the magnetic field intensity, and which has a parallel phase velocity comparable to the thermal velocity of the particles, then the parallel force acts in precisely the same way as the parallel electric field in Landau damping, interacting with resonant particles to transfer energy from the waves to the particles. The power absorbed per unit volume in a uniform plasma can be found from the standard quasi-linear expression for power absorption via Landau damping with the substitution of $k_z B_z$ for the electric field and the average value of $\mu$ for the ion charge (Stix 1962).

In terms of the usual dispersion relation, where the magnetic field is eliminated in favour of the electric field, variations in the parallel magnetic field correspond to variations in the transverse electric field. The power absorbed by this process is given by $\langle J_y E_y \rangle$ (Lashmore-Davies 1972) if it is assumed, as before, that the steady magnetic field is in the $z$ direction and the component of the wave vector perpendicular to the field is in the $x$ direction (Lashmore-Davies interchanges the roles of the $x$ and $y$ axes). To lowest order in the Larmor radius this involves the $\varepsilon_{yy}$ and $\varepsilon_{yz}$ elements of the dielectric tensor. If the imaginary parts of these elements are included in calculating the dispersion relation, the contribution of transit time damping is automatically included.

# 2  Alfven Wave Heating

## 2.1 Introduction

Alfven waves, in the frequency range below the ion cyclotron frequency, are the lowest frequency (typically a few MHz) waves commonly used to heat plasmas. A comprehensive account of the properties of these waves has been given by Hasegawa and Uberoi (1982) and by Cross (1988), in a companion volume in this series, and reference may be made to these for further details and for properties of the waves not directly relevant to heating.

Alfven waves are generally encountered within the context of magnetohydrodynamics, where three wave modes, the Alfven and fast and slow magnetosonic waves occur. For the purposes of tokamak heating, the slow magnetosonic wave plays no role, being generally so strongly damped that it cannot propagate for any appreciable distance into the plasma. In practice only the other two modes occur. In this context these are usually referred to as the fast and slow Alfven waves, following the terminology of cold plasma wave theory. The essential properties of these modes and of their absorption will be discussed in the next section, while subsequent sections go on to discuss the complicating effects of tokamak geometry and relevant experiments.

Since these waves are at a comparatively low frequency, the eikonal approximation is not so useful here as in the higher frequency ranges, and the absorption process is very much related to the inhomogeneity of the plasma. This leads to rather subtle mathematical problems, and a wave spectrum which is dependent on the model used to describe the plasma. In practice, this is probably the least important of the schemes which we shall discuss, since it has been used on only a few tokamaks and appears to give trouble with impurities in most of these. In other frequency regimes it has been shown that problems with impurities can be reduced by careful design of the antenna. A more important problem with Alfven wave heating is that no significant heating has been seen in experiments, there being an apparently poor coupling between the antenna and the plasma. For this reason, and also because Alfven waves are discussed in a companion volume, this chapter is shorter and somewhat less detailed than most of the others.

## 2.2 Properties of the Alfven wave and the heating mechanism

The basic dispersion relation for Alfven waves is obtained from a single fluid description of the plasma in which the electromagnetic force on the plasma is the $J \times B$ force and the displacement current is neglected. The electric and magnetic field are related through Ohm's law

$$E + v \times B = \eta J$$

where $\eta$ is the resistivity, and also the Maxwell equation

$$\nabla \times B = -\partial B/\partial t.$$

We shall consider a perfectly conducting plasma, so that $\eta$ is zero. The effect of its being non-zero would be to include collisional damping of the wave, an unimportant effect in fusion plasmas. The dispersion relation is obtained by linearizing the equations and assuming that all perturbations vary as $\exp(ik \cdot r - i\omega t)$, a standard procedure the details of which we shall not discuss here. It predicts the existence of three wave modes with dispersion relations

$$\omega^2 = V_A^2 k_z^2 \tag{2.1}$$

and

$$\omega^2 = \tfrac{1}{2} \{c_s^2 + V_A^2 \pm [(c_s^2 + V_A^2)^2 - 4c_s^2 V_A^2 \cos^2 \theta]^{1/2}\}. \tag{2.2}$$

In magnetohydrodynamics, the first of these is known as the Alfven wave and the other two as the fast and slow magnetosonic waves. The magnetosonic waves involve compression of the plasma, and so their dispersion relation involves the sound speed $c_s$. The Alfven wave, on the other hand involves only a shear motion with no compression, the restoring force is provided solely by the magnetic field, and the dispersion relation depends only on the Alfven speed, defined by

$$V_A^2 = B^2/\mu_0 \rho$$

with $\rho$ the plasma density.

In a tokamak, the plasma pressure is small compared to the magnetic pressure, and so, as has been mentioned in the introduction, the waves of interest can be derived from cold plasma theory and the essential properties of Alfven wave heating do not depend significantly on finite pressure effects. A possible exception to this would, of course, be the slow magnetosonic wave, which can be seen from (2.2) to disappear entirely if the sound speed goes to zero. However, the parallel phase velocity of this wave is equal to the sound speed of the plasma, and in a typical tokamak plasma, where the ion and electron temperatures are approximately equal, this is of the order of the ion thermal velocity. The result is that this wave is very strongly Landau damped and plays no role in tokamak heating schemes.

In consequence, the waves can be described to a good first approximation by the low frequency limit of the cold plasma dispersion relation. Taking this limit in (1.7), we have

$$\varepsilon_\perp \approx 1 + \frac{\omega_{pe}^2}{\Omega_e^2} + \frac{\omega_{pi}^2}{\Omega_i^2} \approx 1 + \frac{\omega_{pi}^2}{\Omega_i^2}$$

$$\varepsilon_{xy} \approx \frac{1}{\omega} \left( \frac{\omega_{pe}^2}{\Omega_e} + \frac{\omega_{pe}^2}{\Omega_i} \right) = 0$$

$$\varepsilon_\parallel \approx 1 - \frac{\omega_{pe}^2}{\omega^2} \approx - \frac{\omega_{pe}^2}{\omega^2}.$$

The element in the bottom right-hand corner of the determinant in (1.7) is much larger than any of the others (by a factor of the order of $m_i/m_e$), and so a good approximation to the solution may be found by equating the term multiplied by it to zero. This gives the two roots

$$n_z^2 = 1 + \frac{\omega_{pi}^2}{\Omega_i^2}$$

$$n^2 = n_x^2 + n_z^2 = 1 + \frac{\omega_{pi}^2}{\Omega_i^2} \tag{2.3}$$

and if we introduce the Alfven speed

$$V_A = \left( \frac{B_0^2}{\mu_0 \rho} \right)^{1/2}$$

where $\rho \approx n_0 m_i$ is the plasma density, these are equivalent to

$$\omega = k_z V_A$$

and $\tag{2.4}$

$$\omega = k V_A$$

where we have assumed that $V_A \ll c$, as is usually the case.

In this approximation we have obtained two rather simple modes, one of which (the fast or compressional Alfven wave) propagates isotropically with phase and group velocity both equal to $V_A$. The other mode, the slow or shear Alfven wave, has a dispersion relation which depends only on the wavenumber component along the background magnetic field. These results are easily seen to be equivalent to the zero-pressure limit of the magnetohydrodynamics (MHD) results, though the difference in terminology between plasma physicists and workers in MHD should be noted in order to avoid confusion. The group velocity of the shear Alfven wave is along the magnetic field and energy cannot propagate across field lines. This is the feature of this wave which is essential in the heating scheme. In its simplest terms it involves feeding energy into the shear wave. If the parallel wavenumber is fixed, as in the usual tokamak geometry, then such a wave can only propagate on the surface where the magnetic field and plasma density take just the right values for the dispersion relation to be satisfied. Energy is trapped on this surface and cannot propagate away across the field lines. The object is therefore to excite energy in the fast wave, which can propagate across the field. At the point where its parallel wavelength matches that of the slow wave of the same frequency, some of its energy is transferred to the latter. Since this energy cannot escape across the field lines it must be deposited locally. If kinetic effects are taken into account, it is found that energy, to some extent, propagates across the field lines in the kinetic Alfven wave, which we shall discuss later. This affects the details of the energy deposition profile, but does not affect the general features of the absorption process. We shall now see how this physical model translates into a theoretical description of the absorption, and how it is modified if we introduce extra physics into the picture.

To investigate the essential structure of the equations which govern this process we consider a slab model in which the steady magnetic field is in the $z$ direction and the plasma properties vary with $x$. We use the fact that in the cold plasma approximation the electric field obeys the equation

$$\nabla \times (\nabla \times E) = \frac{\omega^2}{c^2}\, \varepsilon \cdot E \qquad (2.5)$$

with $\varepsilon$ the dielectric tensor. With the approximations to the dielectric tensor as used above, and the assumptions that the field goes as

$$E(x)\exp(ik_z z + ik_y y - i\omega t)$$

this gives

$$\frac{d}{dx}(ik_y E_y) + (k_y^2 + k_z^2)E_x = \frac{\omega^2}{V_A^2} E_x$$

$$ik_y \frac{dE_x}{dx} - \frac{d^2 E_y}{dx^2} + k_z^2 E_y = \frac{\omega^2}{V_A^2} E_y$$

where we have taken $E_z = 0$ since, as explained above, it is multiplied by a large parameter in the equation. In the two-fluid model this corresponds to the neglect of electron inertia, while in magnetohydrodynamics it is a straightforward consequence of putting the plasma resistivity equal to zero.

Combining these two equations leads to

$$\frac{d}{dx}\left(\frac{k_z^2 - \omega^2/V_A^2}{k_y^2 + k_z^2 - \omega^2/V_A^2}\frac{dE_y}{dx}\right) - (k_z^2 - \omega^2/V_A^2)E_y = 0. \quad (2.6)$$

Equations of this general form have been considered by Tataronis and Grossmann (1973) and Chen and Hasegawa (1974) in the context of Alfven waves, and were also obtained by Barston (1964) and Sedláček (1971) in studies of electron plasma waves in an inhomogeneous system.

Unless $k_y = 0$, equation (2.4) is singular if

$$k_z^2 = \omega^2/V_A^2$$

at some point in the plasma, since then the coefficient multiplying the second derivative goes to zero. This singularity gives rise to the damping of the Alfven waves when the mathematics of the problem is treated. The relevant mathematics is rather complicated and rather than go through it in detail we shall simply try to convey the main ideas involved and relate them to the physics of the absorption process. A particularly thorough treatment of the mathematics is given in the work of Sedláček referred to above. Many of the details can also be found in the book by Cross (1988).

Physically, the singularity coincides with the layer on which the shear wave can propagate. It is clear from (2.6) that the shear and compressional modes separate out in a homogeneous plasma, and that the coupling between them is entirely a result of the plasma gradient. In the shear wave the velocity perturbation is perpendicular to both the magnetic field and the wave vector. With a non-zero component of $k_y$, the velocity perturbation has a component along the $x$ direction, that is, along the density or magnetic field gradient. This can then produce a coupling to the compressional wave.

We shall suppose that we wish to solve (2.6) in some range of $x$, with appropriate boundary conditions, and that $\omega^2/V_A^2$ varies smoothly across this region. If the coefficients of (2.6) were non-zero and the solutions bounded, we should expect such an equation to give rise to an eigenvalue problem, with a discrete set of values of $\omega$ allowed. With an imposed oscillation from an external source we would then anticipate that there would be resonance and strong absorption at this finite set of frequencies. The singularity of the equation gives rise, however, to quite different behaviour. If the frequency is such that

$$\omega^2 = k_z^2 V_A^2$$

is satisfied anywhere in the plasma, then there is a solution with a logarithmic singularity at this point. This is just the standard behaviour of a differential equation with a regular singularity. To be more specific, let us assume that the variation of $V_A$ with $x$ is monotonic, so that there is only one such point. Then around the singularity we may assume $k_z^2 - \omega^2/V_A^2$ is a linear function of $x$, so that the equation in this neighbourhood takes the form

$$\frac{\mathrm{d}^2 E_y}{\mathrm{d}x^2} - \frac{1}{x}\frac{\mathrm{d}E_y}{\mathrm{d}x} - k_y^2 E_y = 0 \qquad (2.7)$$

the origin having been chosen to be at the singularity. This is a standard equation whose solution is of the form

$$E_y = a I_0(k_y x) + b K_0(k_y x) \qquad (2.8)$$

with $I_0$ and $K_0$ the modified Bessel functions. The $K_0$ function has a logarithmic singularity at $x = 0$, and in continuing the solution through this point we must have recourse to the usual arguments involving either causality, as in Landau damping, or the introduction of a small amount of damping. In either case the result is that we take the limit as the singularity approaches the real axis in the direction corresponding to taking the limit as the imaginary part of $\omega$ tends to zero from above (assuming the time variations to be as $\exp(-\mathrm{i}\omega t)$). From Maxwell's equations it follows that near $x = 0$, $E_x$ and $B_y$ go as $1/x$ while $B_x$ has a logarithmic singularity and $B_z$ is finite. Continuing around the singularity in the complex plane, as described above, produces a discontinuity in the imaginary part of $E_y$ which, in turn, leads to a discontinuity in the Poynting vector, corresponding to absorption of the wave at the resonance. With no driving term, the extra freedom introduced by the presence of the singular part of the solution allows the boundary conditions to be satisfied with any $\omega$ for which the singularity exists in the plasma. There is therefore a continuous spectrum, with singular eigenfunctions, similar to the spectrum for electrostatic oscillations in a hot plasma (Van Kampen 1955, Case 1959). If the system is driven there is a similar continuous resonance spectrum, with a singular response at the point in the plasma where $\omega^2 = k_z^2 V_A^2$.

Such a singular response is not, of course, an acceptable solution in reality. There are two ways within cold plasma theory in which it can be resolved more satisfactorily. One is to treat the problem as an initial value problem, like the standard theory of Landau damping. Starting from a uniform plasma there then appears a spike in the field around the resonance, increasing in amplitude with time. The damping of the wave corresponds to transfer of energy into this localised field, just as Landau damping corresponds to a steady transfer of energy to the resonant particles. The other possibility is to introduce a small amount

of damping into the problem and by so doing move the singularity off
the real axis. The amount of energy dissipated is found to be independent of the damping, since all its presence does is determine the way in
which the solution is continued past the singularity. A more physical
explanation is that if the average rate of dissipation per unit volume
$\frac{1}{2}\boldsymbol{E}\cdot\boldsymbol{J}^*$ is calculated, it is found to go as

$$I_m\left(\frac{1}{x - \mathrm{i}v}\right) = \frac{v}{x^2 + v^2}$$

where $v$ depends on the damping. Integrating this quantity over $x$ gives
$\pi/2$, regardless of the value of $v$. Thus the total dissipation is the same,
but the larger is $v$, the wider is the profile of the dissipation and the
lower is the peak value.

Some further physical insight into the nature of the Alfven wave
absorption can be gained by recalling the nature of the shear Alfven
mode, which is such that the group velocity is along field lines and no
energy propagates across field lines. If the plasma is excited so that $\omega$
and $k_z$ satisfy the shear Alfven wave dispersion relation at some layer in
the plasma, then the wave is resonantly driven around that layer, and
the energy fed into it is trapped in the neighbourbood. This leads to
continuous growth of the wave amplitude around the resonance, unless
it is limited by damping.

Another resolution of the problem of the singularity occurs if we go
beyond the simple fluid theory and take account of finite Larmor radius
effects through a kinetic theory as has been discussed by Hasegawa and
Chen (1976). These authors use the linearised Vlasov equation to
describe the ion response and a drift kinetic equation to describe the
electrons. The latter, valid because the wave frequency is much less than
the electron cyclotron frequency, assumes that the perpendicular motion
of the electrons is given by the $\boldsymbol{E} \times \boldsymbol{B}$ drift. With various further
assumptions, most importantly that the wave frequency is much less
than the ion cyclotron frequency, so that a quasi-neutrality condition can
be imposed, and that the wavelength is greater than the ion Larmor
radius, the equations can be reduced to a single differential equation.
The essential feature of this is that terms involving higher-order derivatives, of up to fourth order, are added to equation (2.6). With these
finite Larmor radius corrections included the coefficient of the highest
derivative of the equation no longer goes to zero when $\omega^2 = k_z^2 V_A^2$, and
so there is no singularity. On the other hand the order of the dispersion
relation is raised, so that there is an extra solution representing a wave
propagating across the field, the so-called kinetic Alfven wave. The
singularity of the fluid theory is replaced by a mode conversion and
energy is carried away by the kinetic Alfven wave towards the high
density side of the resonance layer. This wave, whose wavelength

decreases rapidly away from the resonance, may be damped by Landau or collisional damping. Other physical effects, for example finite electron mass, which are neglected in cold plasma theory may also introduce terms which get rid of the singularity. Electron mass is particularly important near the plasma edge, where the ratio of plasma frequency to wave frequency is smaller.

The total energy extracted from the source is the same as that obtained in fluid theory, the difference being that finite Larmor radius effects give the shear Alfven wave a small component of group velocity across the field lines, so that it is able to carry energy away from the resonance layer and does not tend to become of infinite amplitude in the absence of dissipation. Provided that damping is sufficiently strong to produce total absorption within the plasma, the total energy absorbed is not changed by the kinetic corrections, though the profile of the absorption is altered. A similar phenomenon occurs in the absorption of an electromagnetic wave by resonant absorption in an inhomogeneous unmagnetised plasma. Here energy is fed into a plasma wave which has zero group velocity in the cold plasma approximation, but propagates when finite temperature effects are included. A detailed examination of why absorption remains unchanged in this case has been carried out by Pert (1978).

Hasegawa and Chen (1976) also consider the possibility of non-linear dissipation of the kinetic Alfven wave. A parametric decay process is possible, in which the original kinetic Alfven wave couples to a similar wave of lower frequency and an ion acoustic wave. This non-linear process is enhanced because of the fact that the kinetic Alfven wave has a larger field amplitude than the externally excited wave, in consequence of its small group velocity across the magnetic field. The non-linear damping may enhance the rate of dissipation of the kinetic Alfven wave, but again has no effect on the overall energy dissipation.

## 2.3 Effects of cylindrical and toroidal geometry

The above considerations have been based on a slab geometry, but because the Alfven wave is at the lower end of the frequency range used in radiofrequency heating, it is a scheme in which the effects of the toroidal geometry of the plasma are of fundamental importance. Many of the important effects can be obtained in cylindrical, rather than toroidal, geometry, and since this obviously represents an important simplification, it is with this case that we shall start.

The basic equations describing small-amplitude, low-frequency perturbations in a current-carrying plasma have been given by Appert and

Vaclavik (1983). The theory starts from the linearised momentum equation

$$\rho \partial v / \partial t = (j \times B_0 + j_0 \times B) \qquad (2.9)$$

and a linearised Ohm's law which includes the Hall term, that is the force on the current-carrying particles due to the magnetic field, and takes the form

$$E + (v \times B) = m_i/(e\rho)(j \times B_0 + j_0 \times B). \qquad (2.10)$$

A cylindrical geometry is assumed, with the equilibrium quantities, identified by the subscript zero, dependent only on the radius $r$. Perturbations can then be considered as a superposition of cylindrical modes going as

$$\exp[i(kz + m\theta - \omega t)]$$

where $k$ and $m$ are the axial and azimuthal wavenumbers and any singular layers are resolved, there being no damping in these equations, by taking the limit as $\omega$ tends to the real axis from above. Combining (2.9) and (2.10) with Faraday's law leads to a relation between $j$ and $E$, and hence an expression for the plasma dielectric tensor, following the standard type of procedure. The form of Ohm's law used above, with a Hall term, ensures that corrections of order $\omega/\Omega_i$ are included in the tensor elements. It also implies that the component of electric field parallel to $B_0$ is zero, so that the dispersion tensor reduces to a $2 \times 2$ form. We shall see a similar approximation appear in the next chapter where we consider the ion cyclotron frequency range. It is simply a consequence of the fact that at frequencies well below the electron plasma frequency the electrons can move along magnetic field lines to cancel out any parallel electric field.

The result is substituted into Maxwell's equations, expanded to first order in $B_{0\theta}/B_{0z}$, and a pair of coupled equations for the perpendicular electric field and the parallel magnetic field obtained, of the form

$$A \frac{1}{r} \frac{d}{dr} (rE_\perp) = Gk_\perp E_\perp + \frac{i\omega}{c}(A - k_\perp^2)B_\parallel \qquad (2.11)$$

and

$$A \frac{dB_\parallel}{dr} = i \frac{c}{\omega}(G^2 - A^2)E_\perp - Gk_\perp B_\parallel. \qquad (2.12)$$

The coefficients in these equations are defined by

$$A = (\omega/V_A)^2 \frac{1}{1 - (\omega/\Omega_i)^2} - k_\parallel^2 \qquad (2.13)$$

and

$$G = \left(\frac{\omega}{V_A}\right)^2 \frac{\omega/\Omega_i}{1 - (\omega/\Omega_i)^2} - \frac{2 B_{0\theta}}{rB_{0z}} k_\parallel \qquad (2.14)$$

while the parallel and perpendicular wavenumber components are given by

$$k_\parallel B_0 = kB_{0z} + (m/r)B_{0\theta}$$

$$k_\perp B_0 = (m/r)B_{0z} - kB_{0\theta}.$$

Solutions of (2.11) and (2.12) under various conditions are discussed by Appert and Vaclavik (1983) and Appert *et al* (1984, 1986). Here we give a brief summary of the main conclusions of their work.

A vital feature of the equations is that they have a singularity at any point where $A = 0$, which is clearly the Alfven resonance we have already discussed for slab geometry, with a slight modification coming from finite frequency effects introduced through the Hall term in Ohm's law. The simplest case is that of a homogeneous plasma cylinder surrounded by a vacuum region between the plasma and a perfectly conducting wall. If, in addition, there is no current in the plasma, then $G$ is a constant and the equations can be combined into a Bessel equation for the parallel field perturbation. The Bessel functions which determine the radial variation of the fields are functions of $k_r r$, where

$$k_r^2 = (A^2 - G^2)/A. \qquad (2.15)$$

Solutions must be obtained which satisfy the radial boundary conditions of the problem. If the vacuum region is absent and the plasma is uniform right up to a conducting wall, then the solution must go to zero at the boundary and we need a solution of the form $J_m(k_r r)$, with $k_r$ taking a discrete set of values. For such a solution $k_r^2$ must be greater than zero. The corresponding eigenfrequencies are found from the dispersion relation (2.15), which determines the possible values of $\omega$ once the set of values of $k_r$ satisfying the boundary conditions has been found. The Bessel function $J_m$ has an infinite number of zeros, so the allowed values of $k_r$ form an unbounded sequence to which there corresponds an infinite set of frequencies. From (2.15) it is clear that $k_r \to \infty$ as $A \to 0$, assuming that $G$ is non-zero. To values of $A$ in any arbitrarily small neighbourhood of $A = 0$ there corresponds a range of $k_r$ which contains an infinite number of values satisfying the boundary condition. There are, in consequence, an infinite number of eigenfrequencies arbitrarily close to the frequency given by $A = 0$. This last frequency is an accumulation point of the spectrum, approaching which the frequencies become more and more closely spaced.

The dispersion relation is quadratic in $\omega^2$ so that there are two wave modes which can be identified with the homogeneous plasma Alfven and fast magnetosonic modes. This assumes that $G$ is non-zero, that is, that the finite frequency effects are included, or, if the MHD limit is taken, that the current is non-zero. In the limit $G = 0$, the Alfven wave spectrum becomes degenerate, given by $\omega^2 = k^2 v_A^2$ irrespective of the radial structure. If there is a vacuum region between the plasma and the wall, it is possible to construct solutions with $k_r^2 < 0$, in terms of superpositions of the modified Bessel functions. In the plasma the solution must be proportional to $I_m(k_r r)$, since the second solution of the equation is singular at $r = 0$, while in the vacuum region it can be a combination of the two solutions. Again this is an eigenvalue problem which determines the possible frequencies. The solutions in this case decrease away from the plasma surface and in the limit of a plane interface would correspond to surface waves.

In a real system, of course, we do not have a uniform plasma cylinder, but a diffuse density profile. If there is a surface in the plasma on which $A = 0$, then the equations describing the fields are singular there. This is just the Alfven resonance we have already looked at in plane geometry, with a finite frequency correction. At the radial position of the singularity the equations have a logarithmic singularity, and, just as in the case of slab geometry, the extra freedom introduced by the singular part of the solution allows us to satisfy the boundary conditions for any value of the frequency. Thus we have a continuum of eigenfrequencies, covering the entire range of frequencies for which $A = 0$ at some point in the plasma. The corresponding eigenfunctions are localised around the singular surface and it is necessary to find some scheme which will transmit energy from the surface of the plasma to the resonant layer. For typical tokamak parameters the mode which is most effective in producing absorption at a resonant layer well within the plasma has the character of a surface mode. In other words it is such that the excitation generated at the plasma edge decays into the plasma, the local value of $k_r^2$ being negative, so that in the WKB approximation the wave is evanescent. However, the wave can tunnel through to the resonance layer where it excites the slow Alfven wave which, just as we discussed for slab geometry, is confined to the vicinity of the resonance.

At frequencies below the lower end of the continuum there usually exists a series of discrete eigenfrequencies, corresponding to what are generally called global modes, since they involve a cavity oscillation of the whole system, rather than an excitation localised around a resonant surface. Since there is no resonance they are generally weakly damped and not particularly useful from the point of view of heating. The frequencies form a countably infinite set, with an accumulation point at the lower bound of the continuous spectrum. In general, however, only the lowest radial eigenmode is of significance.

Kinetic effects on the Alfven wave spectrum have been considered by Mahajan (1984). He shows that inclusion of electron parallel dynamics in the treatment of an inhomogeneous current-carrying plasma introduces higher-order derivatives into the equations describing the system, so that there is no longer a singularity at $A = 0$. The result is that the continuous spectrum is replaced by a discrete spectrum, though there remains a qualitative distinction between the modes in the global spectrum and what was formerly the continuous spectrum. The latter still have fields peaked around the layer where $A = 0$, though there is no longer a singularity there, and are much more strongly damped than the global modes. They are also more closely spaced.

The effect of toroidal, rather than cylindrical, geometry has been discussed by Appert *et al* (1982). The conclusion is that the behaviour is not drastically changed. Essentially this is because the main parameter determining the behaviour is the Alfven speed. Although, in a tokamak, the magnetic field strength does not have cylindrical symmetry around the minor axis, the density, which varies over many more orders of magnitude, does, to a good approximation, with the result that the shape of the resonant surfaces and the form of the modes are not too dissimilar to what would be expected from the calculations in a cylindrical system. The frequencies of the global modes, however, may be dependent on the details of the plasma geometry.

The above should give some idea of the complexities associated with the Alfven wave spectrum in a toroidal plasma, and of the way in which its details are dependent on just what model is used to describe the plasma. We have avoided most of the mathematical detail in favour of a broad outline of the ideas involved, but the references given above should be sufficient to lead an interested reader into the literature of the subject.

## 2.4 Antenna–plasma coupling

In addition to the propagation and absorption of the waves in the plasma, a crucial part of any analysis is a calculation of the way in which the energy is coupled from the antenna to the plasma. The objective is to launch energy into the centre of the plasma and avoid, as far as possible, high electric fields localised in the edge region since these are likely to contribute to the problem of impurities from the wall. Except for the case of electron cyclotron heating, most of the energy is in modes which are evanescent in vacuum and in a low density layer near the edge of the plasma. It is necessary for the waves to tunnel through this layer, which for Alfven waves generally occupies the entire plasma between the edge and the resonance layer.

A good deal of theory has been developed to describe the coupling (e.g. Hofmann *et al* 1984, Ross *et al* 1986, Puri 1987). Here we shall try to describe the way in which such calculations are carried out, without getting bogged down in the technical details of the mathematics. Our description is mainly a summary of the work of Ross *et al* (1986). The first step is to Fourier analyse the current density in the antenna, which is assumed known, with respect to the toroidal and poloidal directions, so that each mode just contains a radial dependence. In doing this it is necessary to take account of the feeders bringing current in through the wall of the tokamak, as well as the antenna itself, since they may make a significant contribution to the field which is generated. It is most often assumed that there is a vacuum region between the wall and the plasma edge, and that the antenna lies in this vacuum region.

A simplification can be achieved by using the approximation of cylindrical rather than toroidal geometry, since there are then analytic solutions for the field in the vacuum layer. Since the field is evanescent in the radial direction for the modes of interest, the solution is in terms of modified Bessel functions. With the antenna currents included an inhomogeneous equation is obtained, the current producing the source term. A Green's function solution can be generated from the Bessel function solutions of the homogeneous equations. If the wall is assumed to be perfectly conducting, usually a reasonable approximation, then a boundary condition of zero tangential electric field has to be imposed there. This still leaves unknown constants in the vacuum layer solution, and these have to be determined by the fact that at the plasma edge the solution matches smoothly onto a solution for the fields in the plasma.

Since we have a linear problem, the amplitude of the field in each mode is proportional to the current in the antenna, and the power deposited in the plasma is proportional to the square of the current. The most useful parameter to sum up the behaviour of the antenna and be compared with experiment is its impedance $Z$, defined by

$$\text{power absorbed} = \tfrac{1}{2} ZI^2$$

where $I$ is the amplitude of the alternating current in the antenna. With perfectly conducting walls the power absorbed should be equal to the power delivered to the antenna, and this can be used to check the accuracy of numerical calculations. Numerical codes are needed to calculate the fields in the plasma and match them onto the vacuum fields in order to obtain the impedance.

The assumption that the density of the plasma goes to zero at some finite distance from the wall is something of an idealization, since there is always some plasma in the shadow of a tokamak limiter. Ross *et al* (1986) discuss the modifications of the boundary conditions which are needed to deal with this case. Regardless of the details of the boundary conditions at the wall, they come to the interesting conclusion that the ratios of field components within the plasma are entirely determined by the properties of the plasma and that the structure of the antenna only affects the overall amplitude. On this basis they suggest that particular modes in the plasma cannot be excited preferentially by suitably designed antennae. Puri (1987) finds, however, that compressional and shear modes may be excited independently, and that their relative amplitudes do depend on the antenna geometry. These differences appear to arise from the precise way in which boundary conditions at the wall and matching conditions at the plasma–vacuum boundary are imposed.

## 2.5 Experiments on Alfven wave heating

Some early experiments on Alfven wave heating were carried out in devices other than tokamaks — $\theta$-pinches and stellarators. These experiments, which have been reviewed by Shohet (1978), showed that energy could indeed penetrate into the plasma and be absorbed in the region of the resonances predicted by theory. These experiments also showed a problem which tends to occur with many radiofrequency heating experiments, namely enhanced transport leading to a degradation of the confinement in the machine.

Turning now to tokamaks, an extensive series of experiments on heating in the Alfven wave frequency range has been carried out on the TCA machine in Lausanne, a small tokamak with major radius 0.61 m and the plasma contained in a vacuum vessel of rectangular cross section. More details than we can give here may be found in Collins *et al* (1986) and in Besson *et al* (1986). A vertical elongation of the vessel allowed space for the antennas for Alfven wave excitation to be at the top and bottom of the plasma torus, with a group in each quadrant around the toroidal direction. The antenna elements consisted of curved stainless steel bars, running parallel to the surface of the plasma, coated with titanium nitride. This coating was designed to have a low sputtering yield, with the object of minimising the contamination of the plasma by impurities from the antenna structures. The multiple antenna structure, with the relative phasing of the excitation of the different elements being controllable, allowed the selective excitation of different modes and an evaluation of their relative advantages.

In line with the theoretical ideas outlined in the previous section, the object of the antenna system was to excite a fast (or compressional) wave which has the nature of a surface mode, and would then couple to the shear Alfven wave resonance in the interior of the plasma column. The current elements in the antenna ran in the poloidal direction, and so created a magnetic field in the toroidal direction, almost parallel to the steady tokamak field. The result was to produce a magnetic pressure modulation at the plasma surface, proportional to $B \cdot B_0$, where $B$ is the oscillating field.

Since the antennae were in eight groups, at the top and bottom in each of the four quadrants of the machine, the poloidal and toroidal Fourier components could be altered by changing the relative phasing and amplitude of the currents in the groups. Also the structure of each group, consisting of six conductors, could be altered. For the various different modes which could be excited in this way, measurements were made of the antenna loading as a function of density. Typically these showed a number of narrow resonance peaks embedded in a fairly smooth spectrum, the interpretation being that the peaks represented the discrete modes and the remainder the continuous spectrum. This may appear to contradict our earlier statement that the discrete spectrum occurs at frequencies below the lower end of the continuous spectrum. However, it must be borne in mind that a real antenna does not excite a pure Fourier mode, but a superposition of a number of modes. Thus it is quite possible for a discrete mode with certain toroidal and poloidal wavenumbers to be embedded in the continuous spectrum of a different mode.

Changing the current direction in some of the antenna groups allowed the dominant modes to be changed. It was found that there was very little mixing of toroidal modes, so that discrete resonances which were strongly excited with one configuration could be completely absent with another configuration. A detailed discussion of the effects of different mode structures is given by Collins *et al* (1986). In all cases the antenna loading spectrum was both qualitatively and quantitatively in good agreement with calculations, and the experiments demonstrated the generally correct nature of the theoretical picture of Alfven wave excitation which we have discussed above.

The heating produced in the TCA tokamak by Alfven wave heating has been discussed by Besson *et al* (1986). Early results showed an increase in electron temperature, but also an increase in impurity concentration. The latter could be controlled, to some extent, by putting suitable coatings on the antenna structures and on the limiters, but relatively high impurity concentrations were a feature throughout the series of experiments. A peaked radiated power profile, together with enhanced sawtooth activity, suggested that the power deposition profile was peaked around the centre of the plasma.

Although the antenna loading appears to be well explained, heating results have been poor and difficult to interpret. Recently (Borg *et al* 1989), it has been established that the energy deposition within the plasma is mainly determined by the density rise in the centre of the plasma during the discharges and by Ohmic effects, rather than by Alfven wave resonance within the plasma. In this work it was also demonstrated that there is a substantial parasitic interaction between the plasma and the antenna. This occurs because the antenna is in direct contact with a plasma scrape-off layer in the shadow of the limiter and a current flows from the antenna to the plasma. This is not, of course, meant to happen, the antenna being supposed to couple to the plasma through its electromagnetic field, not by injecting current into it—hence the term parasitic.

Our concentration on one experiment does not mean that it is the only one devoted to heating in this frequency range. It is, however, the one on which the most detailed series of experiments have been carried out so far. Other experiments have also shown agreement between theory and experiment on the excitation of waves and the structure of the mode spectrum. However, there seems to be some doubt as to whether Alfven waves are capable of coupling energy efficiently from the antenna to the central region of the plasma. This is in contrast to the higher frequency schemes described in the following chapters, and probably does much to explain the comparatively small number of experiments using this type of heating, and the fact that it is rarely mentioned in the context of future reactor designs.

# 3 Ion Cyclotron Heating

## 3.1 Introduction

Heating in the ion cyclotron range of frequencies (ICRF) is probably the most common scheme at present, being used on a considerable number of tokamaks, and is certainly the scheme with which the highest powers have been used, up to 16 MW in JET. It has proved to be very successful in raising the plasma temperature, at the expense of some degradation in the confinement time in early experiments. More recently the problems of impurities from the antenna and its associated structures, responsible for much of this degradation, has been reduced by improvements in antenna design and by coating the wall of the tokamak and the antennae with carbon or other low atomic number materials. Since these are light elements, their presence in the plasma is less of a problem than the presence of heavier metals, since impurity radiation goes as the square of the charge on the ions. Some particularly successful experiments on JET have used beryllium as a coating (Tubbing *et al* 1989). This material has the advantage not only of low atomic number but of acting as a gettering agent for oxygen. Its use leads to a reduction of about an order of magnitude in the concentration of oxygen coming off the walls into the plasma.

In a single species plasma, the best absorption takes place at twice the ion cyclotron frequency, for reasons which will be explained shortly. Often, however, the plasma in a tokamak contains more than one ion species and this gives rise to additional absorption processes. One very important case is when there is a small minority component of one species, in which case there may be a strong absorption by the minority ions at their fundamental cyclotron frequency. Because of this possibility of having different sorts of ions at differing relative concentrations, there are a number of possible scenarios for ion cyclotron heating. A good deal of theory has been developed and its general correctness verified in a number of experiments. Besides being reasonably well understood, ion cyclotron heating has the advantage of operating in a frequency range where high power sources are readily available, and also of being capable of being launched from comparatively simple antenna structures within the plasma vacuum vessel.

In this chapter we shall discuss the essential features of the theory of ion cyclotron heating and the various methods which can be used. We shall then go on to discuss the main experimental results and their relation to the theory.

The original heating schemes in this frequency range, to which we refer above, depend on excitation of a fast magnetosonic wave at the plasma edge. However, it is also possible to excite a Bernstein mode at the edge and for it to propagate to the centre of the plasma and be absorbed. This scheme, which has attracted a lot of attention recently, both theoretically and experimentally, will be discussed at the end of the chapter.

## 3.2 Wave propagation in the ion cyclotron range of frequencies

As in the previous chapter, we may elucidate much of the essential physics of the problem using a slab geometry to approximate the toroidal geometry of the tokamak. We shall assume that the inhomogeneity is in the $x$ direction, and that this is also the direction of $k_\perp$. In effect this means that we are modelling a wave propagating in the equatorial plane of the tokamak. To be more general we should include a non-zero $k_y$, but this complicates the algebra without adding anything essential to the conclusions.

In general $k_\perp v_{th}/\Omega_i$ and $k_\parallel v_{th}/\Omega_i$ are both small compared to unity, with the result that the elements of the plasma dielectric tensor, given by (1.4) and (1.5), are close to their cold plasma values except when $\omega$ is close to a cyclotron harmonic. To determine the way in which the wave propagates from the plasma edge to the absorption region the cold plasma equations are quite adequate, although the absorption itself depends on hot plasma effects. Because of the small Larmor radius condition referred to above, the dielectric tensor elements may be calculated by including only the lowest-order term in the expansion of the Bessel functions. In the sum over $n$ only the $n = 0$ and $n = 1$ terms, which contribute to the cold plasma result, and the value of $n$ appropriate to any resonant terms with $\omega \approx n\Omega_i$ need be included.

With the notation of (1.6) for the cold plasma dielectric tensor elements we obtain the following equation for $n_x^2$

$$\varepsilon_\perp n_x^4 - b n_x^2 + c = 0 \tag{3.1}$$

where

$$b = (\varepsilon_\perp + \varepsilon_\parallel)(\varepsilon_\perp - n_\parallel^2) - \varepsilon_{xy}^2$$

$$c = \varepsilon_\parallel[(\varepsilon_\perp - n_\parallel^2) - \varepsilon_{xy}^2].$$

(3.2)

For the slab geometry $n_\parallel$ (or $n_z$) is constant, while $n_x$ is a function of $x$.

Although the individual dielectric tensor elements go to infinity as $\omega \rightarrow \Omega_i$, it is quite easy to verify that the solution of (3.1) is finite, and only goes to infinity when $\varepsilon_\perp \rightarrow 0$. This condition determines the cold plasma resonances for this geometry, and is independent of $n_\parallel$. In a single species plasma the only such resonances are the familiar upper and lower hybrid resonances. The upper hybrid resonance does not fall within the frequency range of interest to us here, and the lower hybrid resonance is only close to the ion cyclotron frequency in the low density region at the plasma edge. We shall return to the behaviour at the edge, but at the moment we consider the centre of the plasma where the lower hybrid frequency is well above the ion cyclotron frequency range. If more than one species is present, then it is indeed possible for there to be a cold plasma resonance in the ion cyclotron range of frequencies. This is a topic which we shall discuss in section 3.4, since absorption in a plasma with more than one ion species is a process of great importance in this frequency range.

The physical reason why there is no resonance at the cyclotron frequency is that at this frequency $E_x = -E_y$, so that the wave is circularly polarised in such a way that $E_+$ vanishes. As was shown in Chapter 1, this is the component which interacts strongly with the ions. In the cold plasma approximation the field at the ion cyclotron frequency rotates in the same direction as the electrons (and vice versa). Any attempt to set up a wave field rotating in the same sense as the ions may be thought of as being similar to trying to set up an electric field in a perfect conductor. The system responds immediately in such a way as to cancel out the field. Absorption at the cyclotron frequency or its harmonics is a warm plasma effect. To describe it, finite Larmor radius corrections to the dielectric tensor elements must be included.

The dispersion relation (3.1) may be simplified by noting that $\varepsilon_\parallel$ contains a term $\omega_{pe}^2/\omega^2$ which, in the frequency range of interest here, is much larger than any of the other terms in the dielectric tensor. Thus one of the roots of (3.1) can be obtained simply by equating the term proportional to $\varepsilon_\parallel$ to zero, giving

$$n_x^2 \approx [(\varepsilon_\perp - n_\parallel^2)^2 - \varepsilon_{xy}^2]/(\varepsilon_\perp - n_\parallel^2).$$

(3.3)

Making this approximation corresponds to taking $E_z = 0$, and results physically from the fact that at low frequencies the electrons can very effectively short out any electric field parallel to the magnetic field. It is equivalent to neglecting electron inertia. The root obtained above corresponds to the fast magnetosonic wave. The Alfven resonance discussed in the previous chapter occurs when the denominator in (3.3) vanishes. For parameters appropriate to ion cyclotron heating it may generally be assumed that there is no Alfven resonance in the central region of the plasma. The fast wave found above is the mode of interest for ion cyclotron heating, but we should also say something about the other solution of the dispersion relation, which is

$$n_x^2 \approx b/\varepsilon_\perp \approx \varepsilon_\parallel(\varepsilon_\perp - n_\parallel^2)/\varepsilon_\perp \approx \varepsilon_\parallel \approx -\omega_{pe}^2/\omega^2.$$

Thus this solution is highly evanescent and may safely be assumed to play no role in the transport of energy in the high density parts of the plasma.

The approximate dispersion relation (3.3) may also be used to illustrate another important property of the fast wave. If we include a $y$ component of the wavenumber, we obtain

$$n_x^2 + n_y^2 = \frac{(\varepsilon_\perp - n_\parallel^2)^2 + \varepsilon_{xy}^2}{\varepsilon_\perp - n_\parallel^2} = \varepsilon_\perp - n_\parallel^2 - \frac{\varepsilon_{xy}^2}{\varepsilon_\perp - n_\parallel^2}.$$

In the slab geometry $n_y$ is constant and the path of a ray in the $x$–$y$ plane is parallel to its phase velocity, as can be seen by applying the usual ray-tracing equations to the dispersion relation corresponding to the above. As the density increases, the value of $n_x^2 + n_y^2$ given by the above equation increases, so that $n_x$ increases and the ray trajectory becomes more nearly parallel to the density gradient. A similar effect, when a more realistic tokamak geometry is considered, leads to a focusing of the rays along the radial direction and towards the central region of the plasma. Since the high-density central region is where it is usually desired that absorption takes place, this is clearly a useful property.

The comments made above, on the approximations which may be made to the dielectric tensor and the waves which may propagate, apply to the central region of the plasma, and may not apply near the plasma edge where the density is very low and the Alfven velocity much higher than in the bulk of the plasma. In this region it is possible to find the cut-offs which occur when

$$\varepsilon_\perp - n_\parallel^2 = \pm\varepsilon_{xy}$$

and also the lower hybrid resonance, though just which of these occur in the plasma depends on the frequency and on the value of $n_\parallel$. Any cut-off or resonance which occurs marks the boundary between a region in which the wave propagates and one in which it is evanescent. However, such regions of evanescence are confined to narrow bands near the edge of the plasma. In general they do not pose any obstacle to the propagation of the wave into the centre of the plasma, but must be taken into account in considering the coupling of the antenna to the plasma. In the same connection, the slow wave, which we have shown to be highly evanescent within the plasma, may be of some consequence near the edge. Here $\varepsilon_\perp - n_\parallel^2$ may change sign if $n_\parallel^2 > 1$, and the slow wave may propagate in a narrow layer. Energy from the antenna coupled to this would be reflected and localised near the plasma boundary, an effect which is avoided as far as possible by screening off the field component with the necessary polarisation so that only the fast wave is excited. We shall return to the topic of antenna–plasma coupling and the properties of the edge region, but for the moment we concentrate on the high density part of the plasma.

We have arrived at the conclusion that the wave of interest for ion cyclotron heating is the fast wave, which can propagate across the field, and is polarised with its electric field perpendicular to the magnetic field. The frequency is too high for the Alfven resonance discussed in Chapter 2 to occur, except perhaps near the edge of the plasma where it does not produce significant absorption. In a single species cold plasma there is no resonance to produce absorption, so we must look at the warm plasma corrections to see the mechanism of ion cyclotron heating in such a plasma.

If we consider first the fundamental $\omega \approx \Omega_i$, then we may expect damping to occur through the $n = 1$ terms of (1.5), since the argument of $Z(\zeta_i)$ is small and so this function has an appreciable imaginary part. Thermal corrections to the cold plasma result can be found by expanding the Bessel functions in (1.5) in power series in $\lambda_i$ and substituting the result in (3.3). If this is done, it will be found that the lowest-order corrections cancel and that the thermal effects are of order $\lambda_i^2$ which is a very small number for the usual parameter range. We may conclude that absorption at the fundamental is weak and that it cannot provide an effective way of heating a single species plasma in a typical tokamak. If, however, we consider the second harmonic ($\omega \approx 2\Omega_i$) the picture changes, since the lowest-order thermal corrections no longer vanish and the damping depends on terms of order $\lambda_i$, with the result that a useful amount of damping can take place. Second harmonic heating is an important mechanism which has attracted a good deal of attention, so we shall devote a separate section to a description of its essential features.

### 3.3 Heating at the second harmonic

For $\omega \approx 2\Omega_i$ we obtain an approximate dispersion relation by including the $n = 2$ term in (1.4). Thermal effects can be neglected anywhere other than in this term and we may use the approximation $I_2(\lambda) \approx \lambda_i^2/8$, so that the relevant dielectric tensor elements are approximately

$$\varepsilon_{11} = \varepsilon_{22} = 1 - \omega_{pi}^2/(3\Omega_i^2) + \tfrac{1}{2}\lambda_i Z_2$$

$$\varepsilon_{12} = -\varepsilon_{21} = -i\omega_{pi}^2/(6\Omega_i^2) + i\omega_{pe}^2/(2\Omega_i\Omega_e) - \tfrac{1}{2}i\lambda_i Z_2.$$

Here

$$Z_2 = Z[(\omega - 2\Omega_i)/(\sqrt{2}k_z v_i)]$$

and

$$\lambda_i = k_\perp^2 v_i^2/\Omega_i^2 = n_\perp^2 v_i^2 \omega^2/(c^2\Omega_i^2) \approx 4n_\perp^2 v_i^2/c^2$$

with $v_i$ the ion thermal velocity, defined by $T_i = m v_i^2$ (with $T_i$ in energy units as before). A word of warning is appropriate here, pointing out that authors are divided between this definition of the thermal velocity and one with $T_i = \tfrac{1}{2}m v_i^2$. Apparent discrepancies of factors $\sqrt{2}$ can very often be traced to this.

The dispersion relation, with parallel electron motion neglected, as explained above, is thus

$$\begin{vmatrix} A - n_\parallel^2 & -iB \\ iB & A - n_\parallel^2 - n_\perp^2 \end{vmatrix} = 0 \tag{3.4}$$

with

$$A = 1 - \omega_{pi}^2/(3\Omega_i^2) + \tfrac{1}{2}n_\perp^2 v_i^2 Z_2/c^2$$

$$B = \omega_{pi}^2/(6\Omega_i^2) - \omega_{pe}^2/(2\Omega_i\Omega_e) + 2n_\perp^2 v_i^2 Z_2/c^2.$$

Expanding this, and neglecting terms of higher than first order in $v_i^2/c^2$, we obtain a quadratic in $n_\perp^2$. The immediate conclusion which we can draw from this is that introducing the thermal correction has produced another wave mode in the plasma in addition to the fast mode which was predicted by the cold plasma approximation. This extra mode is, of course, the Bernstein mode. The dispersion curves for a homogeneous plasma are as shown in figure 3.1, although it should be noted that the plasma dispersion functions in (1.7) have imaginary parts (except in the limit $k_z = 0$) so that cyclotron damping occurs. This is strongest close to the cyclotron harmonic. In an inhomogeneous plasma, where the wave propagates with constant $\omega$, but $\Omega_i$ is varying in space as the magnetic field changes, figure 3.1 translates into the variation of perpendicular wavelength with distance shown in figure 3.2.

This is the type of behaviour which is characteristic of the process of mode conversion discussed in section 1.3. From figure 3.1 it can be seen that the Bernstein mode is backward propagating, i.e. its group velocity is in the opposite direction to its phase velocity, at least so far as motion across the field is concerned. The behaviour to be expected is as follows. If the wave is incident from the high field side, then part of the energy will be transmitted as a fast wave while the rest is mode-converted to the Bernstein mode propagating backwards away from the

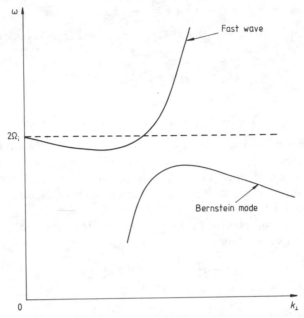

**Figure 3.1** Dispersion curves for the fast mode and the Bernstein mode in the neighbourhood of the second harmonic of the ion cyclotron frequency. In this and in figures 3.2 and 3.3 the real part is shown. If $n_\parallel \neq 0$ there is also an imaginary part resulting from cyclotron damping.

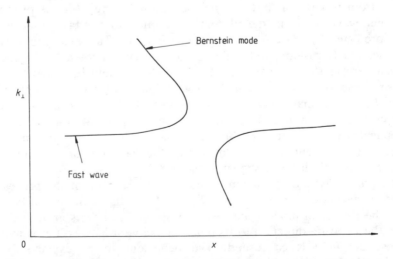

**Figure 3.2** The spatial dependence of wavenumber near the second harmonic. In this and in figure 3.3 the magnetic field is highest at the left-hand side.

second harmonic layer with increasing perpendicular wavenumber. With low field incidence, the wave will again be partly transmitted, but now the mode-converted fraction will go on to the branch of the Bernstein mode which goes to a cut-off. At the cut-off the wave will be reflected and the sign of $k_\perp$ reversed. A further mode conversion will lead to part of the energy going into the reflected fast wave, while part remains in the Bernstein mode propagating towards the high field side. In a tokamak the magnetic field decreases with major radius, so high field incidence corresponds to a wave launched from the inside of the torus, and low field incidence to a wave launched from the outside.

The equation (1.18) was proposed by Ngan and Swanson (1977) in order to investigate wave propagation through this region where it reproduces the main features of the dispersion relation, but does not include cyclotron damping, which is very important in the vicinity of the cyclotron resonance layer. Later work (Swanson 1980, 1981, Colestock and Kashuba 1983) has taken this into account. One problem is finding the correct differential equation to describe the system. In the cold plasma approximation the $k$ in the dispersion relation comes from the term $\nabla \times \nabla \times E$ and so the corresponding differential equation is easily found. However, the warm plasma corrections contain $k_\perp$ in the argument of a Bessel function and the corresponding differential equation is not so easily found. As was pointed out in section 1.3, a simple inverse Fourier transform is not unambiguously defined, and a systematic approximation procedure based on a small Larmor radius expansion of the Vlasov–Maxwell system of equations is required, the result being a fourth-order system of equations describing the coupling of the fast and Bernstein modes.

To give an idea of the kind of calculation involved we shall give an outline of the method used by Colestock and Kashuba, which is based, in turn, on a method proposed by Berk and Dominguez (1977). The basis is a variational principle, which is constructed starting from the Fourier transform of Maxwell's equations which gives

$$k \times (k \times E) = \mu_0 J - i\omega E/c^2.$$

Note that this equation holds, regardless of whether the plasma is homogeneous or not. The properties of the plasma only come in when the relation between $J$ and $E$ is considered.

The inner product in $k$ space is then taken with $E(-k)$, to give

$$\int dk[E(-k)\cdot k \times (k \times E(k)) - \mu_0 E(-k)\cdot J(k) - (i\omega/c^2)E(-k)E(k)] = 0.$$

It is a trivial matter to see that variation of this integral with respect to $E(-k)$ reproduces the original equation.

The next step is to find $J(k)$ in terms of $E(k)$ by the usual procedure of integrating the linearized Vlasov equation along unperturbed trajectories. However, in a system with a gradient along the $x$ axis, say, there is no longer a simple proportionality between these quantities, but instead they are related by a convolution integral so that

$$J(k) = \int dk \; \sigma(k, k')E(k')$$

where $k$ refers to the wavenumber along the direction of the gradient, and we have suppressed the wavenumber components along the direction perpendicular to this as well as the frequency. When this is substituted into the variational principle, the result is a set of coupled integral equations. These simply correspond to Maxwell's equations with the above expression for $J$, in which the plasma inhomogeneity couples different Fourier components of $E$. In order to convert these into differential equations it is assumed that the spatial variation is over sufficiently long scale lengths, so that quantities in which the $k$ dependence comes from these gradients may be expanded to lowest order in $k$ and $k'$. The integral to which the variational principle is applied is converted into $x$ space by inverting the Fourier transforms, and the variational principle is applied to it in this form. The key to the method is, of course, the fact that if, in the $k$ space representation, the fields are multiplied by powers of $k$, then the transformation back to $x$ space gives derivatives of the fields. Truncating the Taylor expansion in $k$ of the slowly varying quantities means that only derivatives up to the corresponding order appear. The algebraic details of this procedure are rather complicated, and we refer the reader to the papers quoted above for further information. Later work by Romero and Scharer (1987) has shown the importance of using a self-consistent equilibrium distribution function with the first-order drifts due to the field gradient included. These add extra terms to the equations which ensure that they conserve energy in the limit when $k_\parallel \rightarrow 0$, in which the wave is undamped. Procedures of this type give a systematic way of deriving the differential equations describing a non-uniform plasma in a form which ensures that energy is conserved in the absence of any damping.

Another problem arises once the equations have been obtained. The Bernstein wave is evanescent on the low field side of the resonance layer and so has exponentially varying solutions there. In the integration, care has to be taken in order to obtain accurate solutions of the fast wave amplitude while avoiding spurious growing solutions of the Bernstein mode amplitude. Swanson, in the references mentioned above, develops a Green's function technique in order to convert the differential equation into an integral equation. The technique involves separating out the terms which produce damping and which involve the plasma dispersion function and putting them on the right-hand side of the equation. What remains on the left-hand side is of a form which, when equated to zero, is amenable to solution using the Laplace transform method outlined in Chapter 1. From the solutions of the homogeneous equation a Green's function can be constructed in such a way as to ensure that the solution has the correct asymptotic properties. Since the full equation contains the unknown solution on the right-hand side, and not just a known driving term, application of the Green's function converts the differential equation into an integral equation, which can be solved numerically. The integral equation is entirely equivalent to the original differential equation, together with the boundary conditions used in constructing the Green's function. Most other authors have used either finite difference (Colestock and Kashuba 1983, Jaeger *et al* 1988) or finite element (Hellsten *et al* 1985) methods. While the problem of obtaining satisfactory numerical solutions is by no means insuperable, it does mean that the calculations are quite time-consuming computationally, and difficult to extend to more than one dimension.

Some recent work on both the second harmonic and the minority heating problems (the latter being discussed in the next section) has used an approach designed to circumvent these problems (Kay *et al* 1986, 1988, Smithe *et al* 1987, Lashmore-Davies *et al* 1988). Recognising that the Bernstein mode is strongly damped near the resonance, except for near-perpendicular propagation, the warm plasma corrections to the dispersion relation are treated as a driven response to the fast wave and in these terms $k_\perp$ is simply replaced by the fast wave value. The only spatial derivative then comes in through the $\nabla \times \nabla \times E$ term and the problem is reduced to a straightforward second-order differential equation. This is done, of course, at the expense of a loss of information about the detailed behaviour of the Bernstein mode. However, comparison with solutions of the fourth-order equations shows that it gives good results for the overall transmission and reflection of the fast wave. For propagation well away from perpendicular, when the Bernstein mode is strongly damped and does not propagate away from the resonance to any significant extent, the absorption profiles are also in good agreement. As the propagation approaches perpendicular, the damping of the Bernstein mode is less strong, and a significant fraction of the incident energy leaves the mode conversion layer in this wave. In general it is expected that this energy will be absorbed elsewhere in the plasma as a result of toroidal effects, which change the parallel wavenumber of the Bernstein mode and lead to its being Landau damped on the electrons (Ram and Bers 1987). The approximation which does not allow for Bernstein mode propagation, on the other hand, absorbs all the energy extracted from the fast wave in the neighbourhood of the cyclotron harmonic resonance. As the damping decreases, the width of the resonance decreases and the absorption profile tends to become a narrower spike. So, although the approximation is generally good as regards the overall energy balance in the fast wave, it is not necessarily reliable as regards the deposition profile of the absorbed energy. Recently some attempts have been made to extend this type of approximation in order to give more exact descriptions of the absorption profile and energy convected away by the Bernstein mode, without going back to a full fourth-order description (Fuchs and Bers 1988, Cairns and Fuchs 1989). In these methods the energy absorbed from the fast wave is calculated by the simplified method described above, but account is then taken of the way in which the Bernstein mode propagates and an estimate made of the way in which it convects energy away from the region in which it is given up by the fast wave.

Another approach, which is entirely analytic, has been developed by Ye and Kaufman (1988). One of the key features of their method is to extend the idea of pairwise coupling of waves in the mode-conversion region, as described in Chapter 1, to propagation in a six-dimensional $(r, k)$ space, in the spirit of ray tracing. The fast waves, which propagate in $r$ space at almost constant $k$, couple to a continuum of Case–Van Kampen modes, representing the dissipative gyroresonant absorption by ions, which propagate in $k$, connecting the incoming fast wave with that going in the opposite direction. Coupling of the incoming fast wave to the Case–Van Kampen modes and thence to the reflected fast mode is considered as a sequence of pairwise couplings and expressions obtained for the transmission, reflection and mode-conversion coefficients, in good agreement with those obtained in other ways.

Since the approximation which treats the warm plasma terms as a driven response does not necessarily require a small Larmor radius expansion in order to produce a differential equation, it has been used by Kay *et al* (1988) to consider the effect on ion cyclotron heating of a small concentration of high energy fusion products. This calculation simply involved evaluating the hot plasma terms with the full Bessel functions rather than the small argument approximation to them. For simplicity the theory assumed Maxwellian distributions for the high energy particles, rather than making any attempt to include non-Maxwellian distributions which are predicted for these. Nevertheless, the results, which indicate strong absorption by a small population of high energy particles, should be qualitatively correct and are of potential interest in a reactor regime.

## 3.4 The two-ion hybrid resonance and minority heating

Although absorption in a single species plasma can take place effectively at the second harmonic, it is common in practice for there to be more than one species present, even if one of them is an impurity which is only present in a small concentration, for example hydrogen in a deuterium plasma. The presence of more than one species allows an alternative heating scheme which is exploited in many of the experiments which have taken place to date. With two ion species a cold plasma resonance appears at a frequency between the two ion cyclotron frequencies (Buchsbaum 1960), and absorption depends on the existence of this resonance combined with the effects of cyclotron damping. We shall begin with a description of the cold plasma resonance, before proceeding to discuss the complications introduced by thermal effects.

The solution for $n_x$, as given by (3.1), has a resonance where $\varepsilon_\perp$ tends to zero, or, in a cold plasma with the ion species labelled 1 and 2, where

$$1 - \omega_{p1}^2/(\omega^2 - \Omega_1^2) - \omega_{p2}^2/(\omega^2 - \Omega_2^2) + \omega_{pe}^2/\Omega_e^2 = 0 \qquad (3.5)$$

(using the fact that $\omega^2 \ll \Omega_e^2$). A graph of the expression on the left-hand side as a function of $\omega^2$ shows that (3.5) is satisfied at some point between the two ion cyclotron frequencies. There is a cut-off close to this resonance, so that in the cold plasma approximation, the propagation in an inhomogeneous system is described by Budden's equation, which is discussed in section 1.2. If warm plasma effects are included, the branch on which $n_x^2 \to \infty$ at the resonance goes over to a propagating Bernstein mode and the dispersion curves look as in figure 3.3. This occurs because the thermal corrections introduce higher-order terms into the dispersion relation, and $\varepsilon_\perp$ is no longer the coefficient of the highest-order term in the equation.

If neither of the ion species has a concentration very much smaller than the other, then the hybrid resonance is not close to either of the cyclotron frequencies and cyclotron damping is weak. Under these circumstances most of the energy which is not transmitted in the fast wave is mode converted to the Bernstein mode, from which it may be absorbed by Landau damping. Even if the $k_z$ of the wave is initially such that Landau damping is very weak, the effect of toroidal geometry is to produce a shift which can lead to strong damping within a short distance of the mode-conversion layer (Ram and Bers 1987). This mode-conversion scheme of heating is most effective when incidence is from the high field side, since low field side incidence can lead to substantial reflection of the incident wave. The energy absorbed from the fast wave can be calculated quite accurately from Budden's formula, even though this neglects Bernstein wave propagation (Jaquinot *et al* 1977). To include Bernstein mode propagation, the techniques outlined in the last section must be used to obtain fourth-order equations.

Many experiments have been carried out in the minority heating regime, where one of the species is present in a small concentration. Under these circumstances, (3.5) is satisfied when $\omega$ is close to the cyclotron frequency of the minority species, something which is quite easy to see if the behaviour of the left-hand side of (3.5) is considered as a function of $\omega$. The minority species makes very little difference over most of the frequency range, but introduces a sharply peaked singularity at the minority cyclotron frequency. In a single species plasma we concluded that there is no strong cyclotron damping at the fundamental, because of the vanishing of the $E_+$ component of the electric field. However, this conclusion is not valid in the case of a minority species, the polarisation of the wave now being effectively dictated by the majority species. The result is that there is a strong interaction between the wave and the minority species, resulting in damping of the wave and heating of the minority species. Early work on this problem ignored mode conversion (Stix 1975), but it was later pointed out that mode conversion to a Bernstein mode was again possible and that the absorption involves a combination of mode conversion and damping.

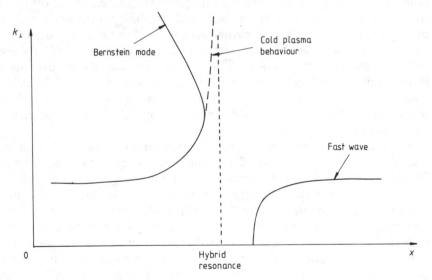

**Figure 3.3** The behaviour of perpendicular wavenumber in the vicinity of the two-ion cyclotron resonance.

Numerical calculation of the absorbed and mode-converted fractions have been carried out by a number of groups (e.g. Colestock and Kashuba 1983, Romero and Scharer 1987) but for the same reasons as were pointed out in the discussion of second harmonic heating, solution of the fourth-order system requires considerable care. The techniques for reduction of the order of the equations, discussed in the last section, have also been applied to this problem (Lashmore-Davies *et al* 1988).

The extra species makes no essential difference to the problem of theoretical calculation of the fate of the incoming wave energy. A common heating scheme involves a minority of hydrogen in a deuterium plasma, in which case there is the added complication that the fundamental resonance for the hydrogen coincides with the second harmonic for the deuterium. In this case there may be significant heating of the deuterium as well as of the hydrogen and both resonant processes have to be taken into account.

## 3.5 Calculations in tokamak geometry and antenna coupling

Although the basic physics can be elucidated using the simple slab geometry, detailed calculations aimed at predicting the behaviour of particular machines have to take account of the toroidal geometry of the plasma. In addition, a realistic calculation needs a detailed consideration of the geometry of the antenna and how it couples to the plasma. The size and shape of the antenna is fundamental in determining the spectrum of parallel wavenumbers which it generates, and this in turn is an important factor in the subsequent propagation and absorption of the wave.

Calculation of wave propagation and absorption in a toroidal plasma presents quite a formidable computational problem which has been tackled in various ways. One possibility is to use ray-tracing techniques (e.g. Koch *et al* 1986), in which the path of the ray is followed using the equations described in Chapter 1, until it reaches the resonant surface on which absorption occurs. The absorption is then modelled by treating the system as having a slab geometry in the vicinity of the resonance and using suitable one-dimensional equations to continue the solution through this region. Although the vacuum wavelength is typically longer than the tokamak dimensions, the wavelength within the plasma is much shorter, and for a larger tokamak it is reasonable to use ray tracing.

Alternatively, the wave equation for the system may be solved in two or three dimensions, using finite difference or finite element methods (Itoh *et al* 1984, Jaeger *et al* 1986, Villard *et al* 1986, Smithe *et al* 1987, Hellsten and Villard 1988). Early codes just took account of the cold plasma two-ion hybrid resonance, with a phenomenological damping included to give energy absorption. At such a resonance the total energy absorbed is independent of the damping coefficient, for reasons discussed in the chapter on Alfven waves, but without a more precise treatment it cannot be expected that the spatial profile of the damping will be correct. Some of the later work (Smithe *et al* 1987) uses the method outlined in the last section, where the hot plasma terms are treated as being driven by the fast wave and are evaluated using the fast wave value for the perpendicular wavelength. Recently a very general computer code containing most of the relevant physics has been developed by Brambilla and Krucken (1988). As well as kinetic damping and the approximate inclusion of the Bernstein mode, they include the effect of a poloidal field component, which converts the usual differential equations into an integro-differential system. Solution of such a set of equations is very expensive in computer time, but provides a very useful validation of simpler methods, with which there is general agreement regarding the overall amount and position of energy deposition. Of course, it also gives information about the behaviour of the Bernstein modes, which is not available from many of the simpler codes.

The other features which must be considered in a realistic calculation are the boundary conditions at the walls and the behaviour of the fields at the antennae which produce the waves. It is generally a reasonable approximation in a tokamak to take the walls to be perfectly conducting, so that the appropriate boundary condition is the vanishing of the tangential component of the electric field at the wall. An antenna for ion cyclotron heating generally consists of a metal structure contained within a Faraday shield, consisting of narrow conducting strips. These strips are aligned along the magnetic field direction and are intended to short out the electric field component in this direction and ensure that the wave is launched with the polarisation of the fast wave. Any energy launched with the other polarisation is trapped near the plasma edge, and may contribute to undesirable effects such as heating the walls and antenna support structures and producing heavy metal impurities.

The simplest treatment of the antenna is simply to treat it as a current sheet with a given distribution of current (e.g. Weynants *et al* 1980, Bers *et al* 1980). This produces a jump in the tangential magnetic field across the antenna, and the equations for the electromagnetic fields in the system have to be solved subject to the existence of this discontinuity. The Faraday shield is generally handled by treating it as a layer with anisotropic conductivity, largest along the direction of the strips. Later calculations have tackled the problem of modelling the three-dimensional shape of the antenna and of calculating the current distribution within it self-consistently with the fields instead of just having an assumed profile (Jacquinot *et al* 1982, Bhatnagar *et al* 1982, Ram and Bers 1984). A variational method for the analysis of antenna–plasma coupling has been developed by Theilhaber and Jacquinot (1984) (see also Theilhaber 1984). Their theory assumes that the wall and antenna are perfectly conducting and that the Faraday screen is perfectly conducting in the direction of the steady magnetic field and has zero conductivity in the perpendicular direction. It is also assumed that there are no cavity eigenmodes, so that the power which tunnels through the layer near the plasma edge, where the wave is evanescent, propagates away into the interior of the plasma and is not reflected back to the antenna. In the mathematics of the problem this means that the boundary condition in the plasma interior requires that there should be a wave propagating away from the wall.

The procedure is to express the electric field as an integral over the current in the antenna, with the full three-dimensional structure incorporating current feeders included. The condition that there be zero tangential magnetic field on the perfectly conducting boundaries produces an integral equation which can be made into an eigenvalue problem, where the antenna impedance is the eigenvalue. The eigenvalue can be calculated from a variational principle. The other papers quoted use varying mathematical techniques, but in all cases the problem is to calculate the fields generated by the current with the appropriate boundary conditions at the wall of the vessel. If it is assumed that there is no reflection from the plasma back to the antenna, then there is the additional requirement of an outgoing wave in the plasma. Calculations of this sort make it clear that the interaction between the antenna and the plasma may depend quite markedly on things like the way in which the current leads are arranged. Careful consideration of such details is needed in order to minimise the wave energy trapped in the edge region. In more recent calculations (e.g. Smithe *et al* 1987), it has been shown that cavity modes can occur because of partial reflection of the wave at the resonance layer. Peaks in the antenna loading occur when there are partially standing waves between the antenna and the resonance layer.

Analysis of the coupling between the plasma and the antenna also requires a knowledge of the behaviour of the plasma waves near the edge, in a region where the approximations we have used earlier may not be valid. From the full dispersion relation, given by (3.1) and (3.2), it can be seen that the nature of the roots changes when one of three things happens.

(i) The constant coefficient $c$ passes through zero. This can occur when $\varepsilon_\parallel = 0$ or when $(\varepsilon_\perp - n_z^2) = \pm\varepsilon_{xy}$. The first of these occurs at a very low density, when $\omega^2 = \omega_{pe}^2$, and the other two, if at all, at slightly higher densities.

(ii) The coefficient of $n_x^4$ goes to zero, so that one root for $k_\perp^2$ goes to infinity. This may occur if $\omega > \Omega_i$, and corresponds to the lower hybrid resonance which, in the present parameter range, is encountered very close to the plasma edge.

(iii) The discriminant of the quadratic vanishes, so that there is a confluence of the two roots. Passing through such a point will take us, in general, from a region where there are two real roots to one where the roots are complex, or vice versa.

An indication of the topology of the dispersion curves near the plasma edge is given in figure 3.4. The lowest density cut-off always occurs, where $\omega^2 = \omega_{pe}^2$, but the others may or may not occur depending on whether the frequency is above or below the ion cyclotron frequency and whether $n_\parallel^2$ is greater or less than one. The condition for these cut-offs is

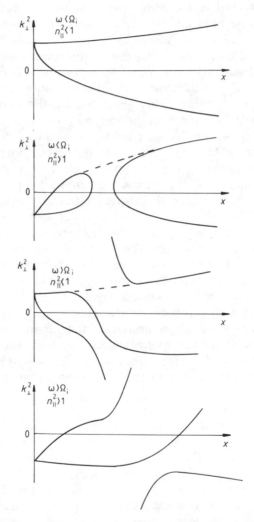

**Figure 3.4** The possible ways in which dispersion curves for ion cyclotron waves may behave near the plasma edge. The broken lines show the behaviour of the fast wave if electron inertia is neglected and the slow wave thereby suppressed. The resonance in the lower two diagrams is the lower hybrid resonance. The positions of the cut-offs are discussed in the text.

$$1 - n_\parallel^2 = \omega_{pi}^2/[\omega(\omega \pm \Omega_i)] - \omega_{pe}^2/[\omega(\omega \pm \Omega_e)] = 0.$$

Approximating the last term on the right-hand side by

$$\pm \omega_{pe}^2/(\omega|\Omega_e|) = \pm\omega_{pi}^2/(\omega|\Omega_i|)$$

we obtain the cut-off condition

$$\omega_{pi}^2 = \mp \Omega_i(\Omega_i \pm \omega)/(1 - n_\parallel^2).$$

If $n_\parallel^2 < 1$ and $\omega < \Omega_i$ there are no positive values of $\omega_{pi}^2$ given by this, while if $n_\parallel^2 < 1$ and $\omega > \Omega_i$ there is one positive value. If $n_\parallel^2 > 1$ there are two positive values if $\omega < \Omega_i$ and one if $\omega > \Omega_i$. Bearing in mind that the lower hybrid resonance is encountered if $\omega > \Omega_i$ and that $(1 - n_\parallel^2)\omega^2/c^2$ is the vacuum value of $k_\perp^2$, it can be deduced that the curves must be of the form shown in figure 3.4.

The most important part of the spectrum is that with $n_\parallel$ greater than one, corresponding to waves which are evanescent in the vacuum. To see this, it is sufficient to note that at typical frequencies of up to the order of 100 MHz, the vacuum wavelength is several metres, and this is much greater than the width of any antenna which will conveniently fit into the machine. Since, for parameters typical of ICRF heating experiments, the detail shown in figure 3.4 is confined to a very narrow layer at the plasma edge, it is doubtful if it has any real significance in practice. Normally, antenna coupling calculations are made using the zero-electron inertia approximation, which eliminates the slow wave and leaves fast wave behaviour shown by the dotted lines in figure 3.4. For $n_\parallel$ greater than one the wave will be evanescent in any vacuum region between the antenna·and the plasma, the width of which will depend on the relative positions of the antenna and the limiters. The width of this region is probably more important than the finer detail near the edge, all of which is squeezed into a distance less than the wavelength in this region.

The calculations of wave propagation and absorption which we have described so far all involve a knowledge of the parameters of the background plasma and generally assume that particle distribution functions are Maxwellian. However, for high power heating the particle distribution function may be modified. If absorption takes place on fast ions and accelerates them to produce a tail on the distribution function, there may be important consequences in the reactor regime. Such a tail can enhance the reaction rate above what would be obtained from a thermal equilibrium distribution with the same energy content.

An analysis of the evolution of the distribution functions under the influence of ion cyclotron heating requires a Fokker–Planck calculation in which the combined effect of the collision term and the wave diffusion term is taken into account, as outlined in Chapter 1. Some early work on this problem was done by Stix (1975) who obtained analytic approximations by expressing the distribution function in terms of $v$, the total velocity, and $\mu$, the pitch angle ($\cos\mu = v_\parallel/v$), and expanding it in the form

$$f(v, \mu) = \sum_{n=1}^{\infty} A_{2n} P_{2n}(\mu)$$

where the $P$'s are Legendre polynomials. Only even orders occur if the wave heating, and hence the distribution function, is assumed to be even in $\mu$. Substituting this into the equation for the distribution function yields a set of coupled ordinary differential equations for the $A$'s. Simplifying assumptions of various sorts can be used to truncate the set of equations and obtain analytic or semi-analytic results. A more recent study which continues this type of work, and gives references to intervening papers, is that of Anderson *et al* (1987). The characteristic feature of these results is that the distribution function is well described by a bulk Maxwellian and a high energy tail. The tail is also Maxwellian in shape, but with a higher temperature than the bulk. The difference between the bulk and tail temperatures, $T_b$ and $T_t$, in the simplest approximation where the tail is assumed to be isotropised by collisions, is given by

$$T_t = T_b(1 + \zeta)$$

where the Stix parameter $\zeta$ is

$$\zeta = P\tau/(3n_i T_b).$$

Here $\tau$ is the collisional slowing-down time for the ions and $P$ is the radiofrequency power input. Essentially, this is just saying that the average extra energy of a particle in the tail is of the order of the rate at which it receives energy from the wave multiplied by its collision time, an intuitively plausible result.

An example of a numerical study of the Fokker–Planck equation applied to ion cyclotron heating is given by Morishita *et al* (1987). Their calculations, and others described by Anderson *et al* (1987), show general agreement with the two-temperature model and the scaling of the tail temperature with the Stix parameter. For minority ion heating, which is discussed by Morishita *et al*, the tail temperature increases as the minority concentration goes down. This is because a low minority concentration involves deposition of the energy in a small number of particles.

An interesting feature of these calculations is that the right-hand circularly polarized component of the wave is found to contribute significantly to the heating. In Chapter 1 we showed that the dominant wave component for ion cyclotron heating is left-hand circularly polarized and that the effect of the right-hand circularly polarized component is smaller, since it is multiplied by a power of the ratio of the ion Larmor radius to the wavelength. However, for particles on the tail, the finite Larmor radius corrections involving the right-hand component may be large enough to make a significant difference to the power absorption (Anderson *et al* 1987).

With increasing computational power becoming available we may expect to see ever more sophisticated and realistic calculations of the behaviour of a tokamak heated with waves in the ion cyclotron range of frequencies. This may be accompanied by further development of simpler approximate methods which, if they can be shown to give results comparable to those obtained with more elaborate calculations, offer a considerable saving in computational resources.

## 3.6 Experiments on ion cyclotron heating

ICRF heating has been used on a considerable number of tokamaks, in both the second harmonic and minority heating regimes. Another possibility, in the same frequency range, is heating by means of ion Bernstein waves excited by an antenna at the plasma edge. The theory and application of this method is discussed in the final section of this chapter. Ion cyclotron heating is probably the most advanced and thoroughly investigated of all the schemes, and is certainly the one for which the highest power to date has been used with up to 16 MW in JET.

The technology required is well developed, with sources being available with output powers of several megawatts in the necessary frequency range. The connection between the generating sources and the antenna is generally by means of coaxial transmission lines, again a reasonably straightforward and well developed technology.

A typical antenna for ICRF heating consists of a metal conductor situated parallel to and close to the wall of the tokamak, as shown schematically in figure 3.5, forming a simple single-turn coil running part way around the minor radius. The elements of the Faraday shield are aligned parallel to the magnetic field, so that the wave transmitted through it is almost entirely polarised with its electric field perpendicular to the magnetic field. This is the polarisation required to excite the fast wave in the plasma. The Faraday screen also helps to shield the antenna from the plasma. Direct contact between the antenna and the plasma may lead to dissipation of power in the immediate neighbourhood of the antenna and a consequent reduction of the coupling efficiency. This effect, the result of currents flowing from the antenna into the plasma, is known as parasitic loading (Colestock 1985).

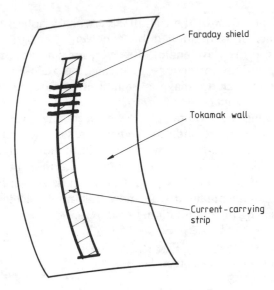

**Figure 3.5** Schematic diagram of an antenna for launching ion cyclotron waves. The long side of the antenna is in the poloidal direction.

While this is the simplest arrangement, others are possible. In JET, for instance, which has used the highest radiofrequency power of any experiment so far, loops are used which have a centre current feed (Jacquinot *et al* 1986). If the two halves are excited so that the current flows in the same direction then this just looks like a conventional (monopole) antenna. If, on the other hand, the two halves are excited with oppositely flowing currents, then the antenna takes the form of a dipole. The effect of this is to suppress the long poloidal wavenumber ($m \approx 0$) modes. There are also antennae consisting of pairs of such loops, with a small separation in the toroidal direction. These are excited in the dipole mode, either in phase, when it just looks like a wider dipole, or in opposition, when they behave as quadrupoles, and suppress the long toroidal as well as poloidal wavenumbers. Such a system can be used to investigate the effect of altering the launch spectrum on the heating efficiency and on the deposition profile. Technical details of a variety of different ICRF antenna designs can be found in articles in Bernabei and Motley (1987). Figure 3.6 gives an idea of the appearance and size of the JET antennae.

An important parameter, which can be compared with theory, is the antenna loading, defined as the ratio of the radiated power to the square of the antenna current. Since the wavenumber spectrum of a typical antenna is mostly in a range where vacuum propagation is impossible, this quantity is very dependent on the separation of the plasma and the antenna. Any change in the plasma configuration which changes this distance will lead to a change in the antenna loading. This is significant if the tokamak makes a transition from the L to the H mode. These two regimes appear to occur in all tokamaks in which auxiliary heating is combined with a divertor configuration. The H mode was first discovered in the ASDEX tokamak at Garching (Keilhacker *et al* 1984) and is characterised by an improved confinement time. From the point of view of the present discussion, the important thing about the L to H mode transition is that it involves a decrease in the edge electron density and an increase in the edge temperature. This produces a decrease in the antenna loading, since the wave has a broader low density region to tunnel through and so less power is transferred to the plasma for a given current. If, as is often the case, there is a feedback mechanism to keep the power input constant, this will lead to an increase in the voltage across the coaxial leads, with the possibility of system breakdown (Steinmetz 1987).

**Figure 3.6** The interior wall of JET, showing two of the ICRF antennae. (Courtesy of JET Joint Undertaking.)

It is clearly more convenient from the technological point of view to have the antenna on the outside of the torus, that is, on the low magnetic field side, rather than on the inside. In the latter case, space may be constricted and access more difficult. In general, the transmission coefficient for the fast wave is independent of the direction of incidence, and the only scheme for which inside launch may be desirable is the mode-conversion scheme, as discussed above, where in some circumstances a substantial fraction of the power may be reflected with launch from the low field side, whereas there is no reflection if the wave is incident from the high field side.

A list of experiments carried out up to 1986 together with their most important parameters is given by Steinmetz (1987). The various methods we have looked at from a theoretical point of view, namely second harmonic heating, mode conversion in a two-species plasma and minority heating, have all been used, and for all of them energy is found to be absorbed, basically as predicted by theory, and to produce an increase in plasma temperature. High power minority heating is found to produce high electron temperatures (Jacquinot *et al* 1987), an effect attributed to the fact that the high energy minority ions lose energy to the electrons much more rapidly than to the majority ion species. This occurs because of the way in which the cross section for Coulomb collisions falls off with the relative velocity of the particles. The minority ions are acclerated to velocities well in excess of the majority ion thermal velocity, so that collisions with the majority ions are not very effective in slowing down the minority ions. On the other hand, the minority ion velocity is still well below the electron thermal velocity, so the ion–electron collision frequency is not reduced.

Steinmetz (1987) characterises various experiments by a dimensionless heating efficiency

$$\eta^* = n_e \, (\Delta T_i + \Delta T_e) V_p / \tau_e P_{rf}$$

where $n_e$ is the average plasma density, $V_p$ the plasma volume, $\Delta T_e$ and $\Delta T_i$ the temperature changes, $\tau_e$ a nominal energy containment time for the experiment and $P_{rf}$ the radiofrequency power. This gives a rough measure of the extent to which the absorbed radiofrequency power goes into the particles and remains in the plasma for the expected confinement time of the machine. The value of this parameter was found to be approximately the same for different machines and for different heating methods, including ion Bernstein wave heating which we discuss in the next section. The conclusion is that no one of the possible ways of heating the plasma using this range of radiofrequency waves is clearly superior to the others.

Early experiments on heating in the ion cyclotron range of frequencies were limited in effectiveness by an influx of heavy metal impurities from the region of the antenna, possibly related to the acceleration of ions in the high radiofrequency fields close to the antenna and the increased sputtering produced by the interaction of these ions with the wall (see, for example, the review by Adam 1984). Such heavy metal impurities cause a drastic increase in plasma radiation losses and may even lead to disruption of the plasma discharge. However, such effects have been mitigated in recent experiments by the use of more suitable materials for the antenna and the shielding of adjacent surfaces with carbon or boron wherever possible. Impurity influx does remain an area of concern, nevertheless, and further studies of the interactions between the plasma, the antenna and the wall are being carried out in order to understand in more detail the conditions under which impurity influx does not cause severe problems. The recent experiments in JET with beryllium-coated walls have achieved a very low impurity level.

Since success in reaching the conditions necessary for nuclear fusion depends not only on reaching a high enough temperature, but also on achieving sufficiently good containment, a great deal of effort has been devoted to the study of the effects of radiofrequency heating on plasma confinement. The rise of stored plasma energy after the heating is switched on is well fitted by the formula

$$W = W_0 + P_{rf}\tau_{in}[1 - \exp(-t/\tau_{in})] \tag{3.6}$$

where $W_0$ is the initial stored energy produced by Ohmic heating. After the radiofrequency heating is switched off, the stored energy decays according to

$$W = W_0 + P_{rf}\tau_{in} \exp(-t/\tau_{in}). \tag{3.7}$$

An example of the way in which observed data fit these formulae is given for JET by Jacquinot *et al* (1986). The characteristic time $\tau_{in}$ which appears here, known as the incremental confinement time, is just the confinement time of the extra energy added by the radiofrequency heating, as can be seen by noting that (3.6) is the solution of

$$d(W - W_0)/dt = -(W - W_0)/\tau_{in} + P_{rf}.$$

while (3.7) is the solution of

$$d(W - W_0)/dt = -(W - W_0)/\tau_{in}$$

with the initial condition $W - W_0 = \tau_{in}P_{rf}$. The incremental confinement time is found experimentally to be less than the confinement time for Ohmic heating, indicating that confinement is degraded as energy is added to the plasma. This appears to be a universal phenomenon, with increased Ohmic heating or neutral beam injection also producing the same result. An empirical law suggested by Kaye and Goldston (1985), on the basis of an analysis of all available tokamak results, has the confinement time varying as approximately the inverse square root of the total power input to the plasma. Although this is a reasonable fit to most radiofrequency heating data, there is some evidence that a somewhat more optimistic scaling with $\tau_E$ proportional to $a + b/P_{tot}$ with $a$ and $b$ constants is a better description (Hosea *et al* 1985, Jacquinot *et al* 1986). This latter law would suggest that at sufficiently high heating power the confinement time should level off at some minimum value, rather than continue to decrease indefinitely. Obviously the way in which the confinement time varies with input power is of the utmost importance, and one of the main aims of higher power experiments will be to settle this question.

Another important result which has been obtained in JET and other machines with high power radiofrequency heating is that sawtooth oscillations appear to be suppressed (Start *et al* 1987). Sawteeth are relaxation oscillations which appear in the central region of a tokamak. The temperature increases, roughly linearly with time, then collapses suddenly before starting to increase again, producing the characteristic graph of temperature against time which gives the oscillations their name (see e.g. Wesson 1987). With sufficiently high power in the ion cyclotron frequency range, it is found that the peak temperature can be maintained for much longer periods, of up to 1.6 s in the results of Start *et al* (1987), with brief temperature crashes between these periods. During the periods between crashes of these so-called 'monster saw-teeth' there is an improved energy confinement time and a very low level of MHD activity. During the temperature crash a strong $n = 2$, $m = 3$ MHD oscillation is set up and may persist for some time. The reason for this behaviour is not well understood at present. Since the collapse of the central temperature redistributes energy within the tokamak and limits the central temperature, the possibility of its suppression is important. The increases in density and temperature associated with the monster sawteeth could be important in enhancing the fusion yield and facilitating ignition in a reactor plasma.

Recently it has been demonstrated in JET that the H mode can be produced using ICRH alone with no neutral beam heating (Tubbing *et al* 1989). This is a very important result, since the energy confinement time increases by a factor of about two in the H mode as compared to the L mode. The initial demonstration of the H mode was in a tokamak heated by neutral beam injection and it was later shown that it could be reached with a combination of neutral beam and ion cyclotron heating. However, there are difficulties in obtaining energy absorption in the centre of reactor plasmas with neutral beam injection, since there is anticipated to be more absorption near the plasma edge than one would wish. The fact that neutral beam heating is not a necessary condition for producing the H mode is thus a very significant result.

The energy confinement found in these experiments was of the same order as that produced by neutral beam injection and the H mode was sustained for periods in excess of 1.5 s. A very important part of these experiments was the use of beryllium as a coating for the walls in place of carbon. Previous experiments using a combination of neutral beam and ion cyclotron heating had shown that there was an increased flux of impurities from the walls and the antenna during the H mode phase, and that the build-up of impurities in the plasma limited the duration of the H mode. This problem was essentially eliminated with the beryllium coating. The main advantage of beryllium is not that it has a lower atomic number than carbon, but that it is very efficient in gettering the oxygen in the walls and leads to a dramatic reduction in the oxygen impurity level.

In general, results with heating in the ion cyclotron range of frequencies may be regarded as very encouraging. The method has proved capable of heating plasmas to temperatures of the sort needed for fusion, without drastic degradation of the energy confinement time, and earlier problems with impurities seem to have largely been solved by improved antenna design and the use of beryllium in recent experiments. A feature of minority ion heating, in particular, is that it can produce a non-Maxwellian distribution with an enhanced population in a high energy tail. In a reactor this could be used to enhance the reaction rate.

## 3.7 Ion Bernstein wave heating

So far we have concentrated on heating by the fast wave, with Bernstein waves only appearing in the vicinity of the cyclotron resonance by virtue of mode-conversion processes. An alternative scheme, however, is one which excites the ion Bernstein wave near the plasma edge and allows it to propagate to the centre of the plasma and be absorbed there (Puri 1979, Ono 1980, 1982). To understand the relevant features of wave propagation we return first to the cold plasma dispersion relation. It is a familiar feature of this that for the extraordinary mode propagating across the magnetic field there is a resonance at the lower hybrid frequency given by (for a single ion species plasma)

$$\omega^2 = (\Omega_e^2 \Omega_i^2 + \omega_{pi}^2 \Omega_e^2)/(\omega_{pe}^2 + \Omega_e^2). \tag{3.8}$$

As the density tends to zero, the lower hybrid frequency tends to $\Omega_i$, while at the centre of a tokamak $\omega_{pe}$ and $\Omega_e$ are generally comparable so that it is of the order of $(\Omega_e \Omega_i)^{1/2}$. Since we are interested in frequencies of the order of $\Omega_i$, the lower hybrid resonance occurs very close to the edge of the plasma, at a density several orders of magnitude below the central density.

The regime of interest here is that with $n_\parallel$ large, in which case the wavenumber does not tend to infinity through real values as $\omega \to \omega_{LH}$ from below, as in the more familiar case of $n_\parallel = 0$, but as $\omega \to \omega_{LH}$ from above. The dispersion curve is thus as shown in figure 3.7 and in the region where $\omega > \omega_{LH}$ (i.e. on the low density side of the hybrid resonance) two waves propagate. The one with smaller $k_\perp$ is the fast wave, which connects smoothly with the fast magnetosonic wave, which we have already considered, in the region where $\omega < \omega_{LH}$. In the present context it is the other wave, the slow wave, which interests us. The terms fast and slow refer to the relative phase velocities of the two waves. The slow wave is backward propagating, that is, its group velocity across the magnetic field lines is in the opposite direction to its phase velocity. According to cold plasma theory, we would not expect it to propagate beyond the lower hybrid resonance, which is very close to the plasma edge, and certainly not where we want the wave to be absorbed. However, if we include thermal corrections, we introduce terms of higher order in $n_x^2$ into equation (3.1) and vanishing of $\varepsilon_\perp$ no longer implies the existence of a resonance. What happens instead is that the slow wave connects to an ion Bernstein mode (Puri 1979) which for large $k_x$ approaches the ion cyclotron harmonic immediately below the lower hybrid resonance, as in figure 3.7. Cyclotron damping of the Bernstein mode is very effective when it is close to the cyclotron harmonic. Thus, if the slow wave is excited at the plasma edge, it changes to a Bernstein mode, which is almost entirely electrostatic in character, in the vicinity of the lower hybrid resonance. The Bernstein mode can then propagate towards the centre of the plasma and be absorbed there. According to linear theory the damping of the wave will only be appreciable in a region where the wave frequency is close to a harmonic of the ion cyclotron frequency. However it has been shown by Abe *et al* (1984) and Porkolab (1985) that a non-linear process is possible which will produce strong damping in a region where

$$\omega \approx (n + \tfrac{1}{2})\Omega_i.$$

The mechanism involved in this absorption is non-linear Landau damping resulting from the self-interaction of the wave (Rosenbluth *et al* 1969). In essence, non-linear effects generate an oscillation at twice the wave frequency, and if this coincides with a harmonic of the ion cyclotron frequency, then this oscillation can interact strongly with the particles and be damped, with consequent heating of the ions. The relevant harmonic must be odd, since otherwise the wave frequency would be at a harmonic of the cyclotron frequency, and this is not allowed by the dispersion relation (Rosenbluth *et al* 1969). Porkolab

(1985) shows that the wave amplitude is governed by an equation of the form

$$\partial E/\partial x + \alpha E + Q(x)|E|^2\, E = 0$$

where $\alpha$ is the linear damping coefficient. The non-linear damping coefficient, proportional to the square of the wave amplitude, is estimated by Porkolab and shown to be large enough, for power levels of interest for heating tokamaks, to ensure that the wave is almost completely absorbed.

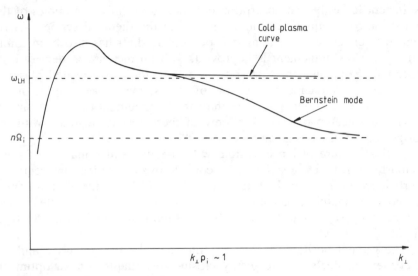

**Figure 3.7** Dispersion curves showing how the slow mode connects to an ion Bernstein wave, for $n_{\parallel}^2 > 1$. The curve deviates from the cold plasma behaviour when $k_{\perp}\rho_i$ is of order 1.

Because the waves of interest have very large values of $n_x$, our earlier arguments concerning the vanishing of $E_z$ no longer hold, and these waves may have a substantial $E_z$ component, particularly near the edge of the plasma where $\omega_{pe}^2/\omega^2$ is not too large. Once this parameter becomes very large, the parallel component of the field becomes small and the $E_x$ component is dominant. To excite the slow wave at the plasma edge, in preference to the fast wave, the Faraday shield on the antenna should be aligned so as to transmit the parallel electric field component.

Early experiments on this method were carried out with detailed diagnostics of the wave, using probes placed at various axial positions in the plasma. These produced data on the radial variation of the wave amplitude and demonstrated clearly that the Bernstein wave had been excited and had propagated into the centre of the plasma in accord with theoretical expectations (Ono 1980). Ono *et al* (1983) give a clear discussion of the theory of linear wave propagation and the way in which the cold plasma wave at the plasma edge changes into the Bernstein mode, and also discuss experimental results which are in good agreement with the theory. They show that at higher harmonics the ion temperature must be increased in order to obtain a smooth transition through the lower hybrid layer.

Subsequent experiments have been carried out at higher powers, sufficient to verify that absorption occurs at half-integer multiples of the ion cyclotron frequency, in agreement with the theory described above, and that effective ion heating can be produced. It has also been found that particle confinement is improved, with an increase in density by a factor of more than three in some ion Bernstein wave heated plasmas. The energy confinement time also increases, mainly as a result of the increased density, but is still less than the confinement time for a similar Ohmically heated plasma. A review of these results, together with the original references, is given by Ono (1987).

A final feature of ion Bernstein wave heating which may be of great importance is that it appears to be capable of suppressing low frequency turbulence in a tokamak. Experimental results from the PLT tokamak demonstrating this effect are described by Ono (1987) and theoretical studies have suggested the possibility of suppressing various modes (e.g. Tripathi *et al* 1987).

The present position on ICRF heating, both by the fast wave and the Bernstein mode, seems to justify closing this chapter on an optimistic note. Absorption appears to take place in the ways predicted by theory, even if some of the finer details of the experiments have not been explained, and to produce heating of the plasma effectively. Although earlier experiments tended to be plagued by impurity problems, with a consequent loss of energy confinement, these have largely been overcome in the most recent experiments.

# 4 Lower Hybrid Heating

## 4.1 Introduction

The lower hybrid frequency is familiar as one of the resonant frequencies of a cold plasma wave propagating at right angles to a magnetic field, and is given by

$$\omega_{LH}^2 = \frac{\Omega_e^2 \Omega_i^2 + \omega_{pi}^2 \Omega_e^2}{\omega_{pe}^2 + \Omega_e^2}. \tag{4.1}$$

In the plasma at the centre of a tokamak $\omega_{pe}^2$ and $\Omega_e^2$ are generally of the same order of magnitude, in which case

$$\Omega_i \ll \omega_{LH} \ll \Omega_e. \tag{4.2}$$

The lower hybrid heating scheme uses waves at a frequency comparable to the value of the lower hybrid frequency at the centre of the plasma, though it is not necessarily the case that the cold plasma resonance plays any role in the absorption. In fact, the frequency used in many experiments is above the lower hybrid frequency everywhere in the plasma, and absorption depends on electron Landau damping. The essential thing is that in this frequency range, typically a few GHz, it is possible for a wave to propagate from the plasma edge to the centre where efficient absorption mechanisms exist. A review of the basic theory, with further references to early work on the problem, is given by Bers (1978).

Lower hybrid waves have been successfully used for plasma heating and current drive. Of all the heating schemes it has, so far, been by far the most successful so far as the latter aspect is concerned. In the regime where electron Landau damping occurs, a long tail on the parallel distribution function is produced which can carry a very substantial current; enough to sustain a tokamak discharge with no Ohmic current drive. This is an aspect to which we shall return in Chapter 6, where we discuss radiofrequency current drive in more detail. The

71

present chapter will give a discussion of the basic propagation and absorption properties of lower hybrid waves and their use in plasma heating, leaving the more specialised aspects of current drive till later.

In this frequency range and with values of the parallel refractive index greater than unity, two waves can propagate, the fast and the slow modes. So far heating and current drive have been obtained most successfully by launching the slow mode, but in a plasma of the size and temperature of a reactor it is expected that this wave would not propagate beyond the edge of the plasma. As a result there has been considerable interest in the possibility of launching the fast mode which, although less effectively absorbed in today's machines, should be able to penetrate a reactor plasma and be sufficiently strongly absorbed near its centre. This aspect of heating in the lower hybrid frequency range is discussed further in the last section of this chapter.

## 4.2 Lower hybrid waves in a cold plasma and the accessibility condition

We begin with a consideration of the properties of the cold plasma dispersion relation in the frequency range well above the ion cyclotron frequency, but well below the electron cyclotron frequency, in accordance with condition (4.2). This will enable us to discuss one of the most fundamental properties of the lower hybrid heating scheme, namely the existence of a critical value of the parallel refractive index below which the wave is trapped near the plasma edge and cannot penetrate to the interior which is, of course, where we would like the energy to be absorbed (Golant 1972, Brambilla 1979, Bonoli and Ott 1982). As is usually the case, the question of accessibility of the wave and its propagation near the plasma edge is adequately described by cold plasma theory, hot plasma effects only being important in describing the absorption processes in the central region of the tokamak.

Again we approximate the essential features of the tokamak geometry by a slab model, so that $n_x$, the perpendicular component of the plasma refractive index, is given by equations (3.1) and (3.2). Now, however, we are in a different frequency range, and the appropriate approximations to the plasma dielectric tensor elements are

$$\varepsilon_\perp \approx 1 + \frac{\omega_{pe}^2}{\Omega_e^2} - \frac{\omega_{pi}^2}{\omega^2}$$

$$\varepsilon_{xy} \approx \frac{\omega_{pe}^2}{\omega \Omega_e}$$

$$\varepsilon_\parallel \approx 1 - \frac{\omega_{pe}^2}{\omega^2}.$$

It is convenient to introduce the notation

$$X = \omega_{pe}^2/\omega^2$$

$$y^2 = \omega^2/|\Omega_e \Omega_i|$$

$$\mu = m_e/m_i$$

in which case (3.1) becomes

$$An_x^4 - Bn_x^2 + C = 0 \qquad (4.3)$$

with

$$A = 1 - \mu X(1 - y^2)$$

$$B = -X[1 - n_\parallel^2 - \mu X(1 - 2y^2)]$$

$$C = \mu y^2 X^3.$$

The wave is excited with $n_\parallel^2 > 1$, so that between the plasma edge and the lower hybrid resonance, $A$, $B$ and $C$ are all positive. Thus, if the roots of the equation are real, they are both positive and two waves propagate. The larger of the two roots gives the slow wave and the smaller the fast wave. This conclusion does not hold right at the edge of the plasma, where the approximations which we have made break down. We shall return to this question shortly, but for the moment references to the plasma edge may be taken to exclude a narrow low density layer next to the plasma wall, in which our assumption that the wave frequency is much less than the electron plasma frequency is invalid.

As a first approximation, it is reasonable to suppose that the parameter $y$, which depends on the magnetic field, is approximately constant as we move across the minor cross section of a tokamak, since it changes much more slowly than $X$, which is a measure of the plasma density. We can then regard $X$ as a spatial coordinate, increasing from the plasma edge to the centre. For small $X$ the roots of (4.3) are real, so the question we must ask is whether they remain real as $X$ increases up to its value at the lower hybrid resonance, where $A = 0$. If there is a change from real to complex roots, it must occur at a point where the roots are equal, the condition for which is

$$\mu^2 X^2 - 2[n_\parallel^2(2_y^2 - 1) + 1]\mu X + (1 - n_\parallel^2)^2 = 0. \qquad (4.4)$$

This is a quadratic in $X$, determining the positions at which there is a transition between real and complex values of $n_x^2$. It has real roots if and only if

$$4\mu^2[n_\parallel^2(2y^2 - 1) + 1]^2 - 4\mu^2(1 - n_\parallel^2)^2 \geq 0$$

or

$$n_\parallel^2 \leq n_c^2 = \frac{1}{1 - y^2}.$$

From this we may conclude that if $n_\parallel^2 < n_c^2$, there is some region between the positions determined by the two real roots of (4.4) in which $n_x^2$ becomes complex and the waves are evanescent. Thus $n_c$ is a critical value of the parallel refractive index and only the part of the spectrum with parallel wavenumber greater than this can reach the central part of the plasma.

To complete our picture of cold plasma wave propagation in the lower hybrid frequency range, we now consider the edge region, where the approximations to the dielectric tensor elements, which we have made in the above, are no longer valid. In the vacuum the solution is

$$n_x^2 = 1 - n_\parallel^2 < 0$$

so the wave is evanescent. There are two cut-offs near the edge, one where $\varepsilon = 0$, i.e. $\omega^2 \approx \omega_{pe}^2$ and another where

$$(\varepsilon_\perp - n_\parallel^2)^2 = \varepsilon_{xy}^2$$

the latter being at a slightly higher density. The first of these is the slow mode cut-off and the second the fast mode cut-off. We can then conclude that the behaviour of $n_x^2$ is summed up in figure 4.1.

The behaviour shown in figure 4.1 may be easier to understand if it is related to the usual $\omega$ versus $k$ dispersion diagram. For $n_\parallel^2 > 1$, the lower hybrid resonance is approached from above, as shown in figure 4.2. In the outer region of the plasma, the wave frequency is above the lower hybrid frequency and, as can be seen from the figure, there are either two modes with oppositely directed group velocities, or, if $\omega$ is above the maximum of the curve, no propagating modes. If $n_\parallel^2 < n_c^2$, then as we go into the plasma with a fixed value of $\omega$, the maximum of the curve in figure 4.2 is first above $\omega$, so that there are two real solutions of the dispersion relation, then drops below $\omega$, so that there are no real roots, then rises above $\omega$ again. When $n_\parallel^2$ is above the critical value, the maximum of the curve never falls below $\omega$ and there are two real roots right up to the lower hybrid resonance. The root with the larger refractive index, which corresponds to the slow wave, gives a backward propagating wave, that is, one with its group velocity across the magnetic field in the opposite direction to its phase velocity.

The usual lower hybrid heating scheme uses the slow wave, though some work has been done on the possibility of using the fast wave, and we shall return to this topic later. As we have seen, the fast and slow waves have oppositely directed group velocities so that if the slow wave is excited with a value of $n_\parallel$ which is less than the critical value, it reaches the point where there is a confluence with the fast wave and propagates back towards the plasma edge on this latter branch. From our earlier discussion of mode-conversion processes (see Chapter 1) it might be thought that some of the energy in the slow wave would tunnel through the evanescent region and continue on the propagating branch of the slow wave at higher density. However, it can be shown (Woods *et al* 1986) that although this is possible in principle, the transmission coefficient changes very rapidly from essentially zero to essentially one as the parallel refractive index goes through its critical value. Only in an extremely narrow range of the spectrum is partial transmission and partial reflection a possibility.

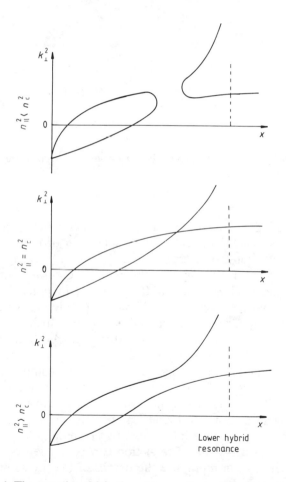

**Figure 4.1** The way in which the wavenumbers of the fast and slow waves vary between the plasma edge and the lower hybrid resonance for different values of the parallel refractive index. The upper curve is the slow wave in all cases.

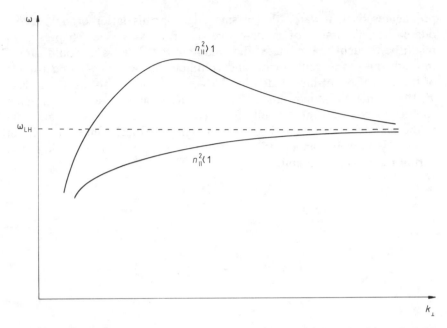

**Figure 4.2** When $n^2 > 1$, the cold plasma curve approaches the lower hybrid frequency from above.

An interesting feature of the mode conversion between the fast and slow waves, is that the directions of propagation of the two waves away from the mode-conversion region are just those which occur in the classical problem of transmission through a potential barrier (see, for example, Budden (1985) section 15.9). Thus, in place of the coupled differential equations described in Chapter 1, this problem is described more naturally by a single differential equation, which yields the energy conservation relation appropriate to this configuration. It is also of interest because we are dealing with cold plasma waves, for which exact differential equations can easily be written down and solved. Thus it provides a good test of the accuracy of the order reduction schemes discussed in Chapter 1. A full discussion of this aspect of lower hybrid wave propagation can be found in Woods *et al* (1986). Using a modification of a technique introduced by Heading (1961), these authors demonstrate that a systematic reduction can be carried out and that the reduced equations can reproduce the details of the full solution to a high degree of accuracy.

Some further properties of lower hybrid wave propagation can be seen if we return to equation (4.3) and note that because $X$ is large and $\mu$ is small, the coefficients $B$ and $C$ are much larger than $A$. The slow wave solution, corresponding to the larger of the two roots, is thus well approximated by

$$n_x^2 = B/A. \tag{4.5}$$

If $n_\parallel^2$ is substantially greater than one, then $B \approx n_\parallel^2 X$, and (4.5) leads to an approximate dispersion relation of the form

$$D = ak_x^2 - bk_\parallel^2 = 0$$

with $a$ and $b$ constants. The perpendicular and parallel group velocities are in the ratio

$$\frac{v_{g\perp}}{v_{g\parallel}} = -\frac{ak_x}{bk_\parallel} = -\frac{k_\parallel}{k_x} = \pm \left(\frac{b}{a}\right)^{1/2}. \tag{4.6}$$

From (4.6) we can deduce that the direction of the group velocity is perpendicular to the wavenumber vector and that the ratio of the perpendicular to the parallel component of the group velocity is independent of $k_\parallel$. This means that if a lower hybrid wave is excited at a point in a uniform plasma, with a spread of parallel wavenumbers, energy does not spread out in all directions, but instead is concentrated into resonance cones around the field direction, as indicated in figure 4.3. Generally $k_x > k_\parallel$ so that these cones are at a small angle to the field.

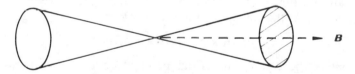

**Figure 4.3** In a uniform plasma the group velocities of lower hybrid waves lie on a resonance cone as illustrated.

Applied to propagation in a tokamak, these results mean that if a spectrum of $k_\parallel$ is excited at the plasma edge, then energy travels into the plasma along well collimated beams, which are mainly in the toroidal direction, in the fashion sketched in figure 4.4. Since there is a large toroidal component in the group velocity, the wave can travel a considerable distance around the tokamak before being absorbed. As the wave travels around in the toroidal direction its value of $n_\parallel$ may change considerably, so that a wave whose initial value of $n_\parallel$ does not lie in the correct range to satisfy the accessibility criterion may eventually penetrate to the centre of the plasma. Examples of ray-tracing calculations showing these effects are given by Colestock and Kulp (1980), Ignat (1981) and Brambilla (1982). Although the slab geometry gives a qualitative account of the accessibility condition, the features of lower hybrid wave propagation discussed here make it essential to use such ray-tracing techniques if an accurate treatment of a toroidal plasma is to be obtained.

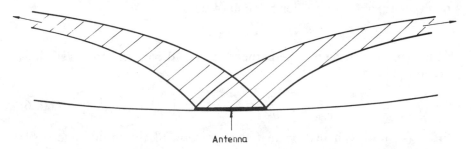

Antenna

**Figure 4.4** When waves are launched from the edge of a tokamak, the behaviour shown in figure 4.3 leads to focusing of the waves into beams propagating mainly in the toroidal direction.

An interesting feature of toroidal geometry is the possibility of ergodic ray trajectories (Bonoli and Ott 1982). The ray-tracing equations are written in terms of the toroidal coordinates $(r, \theta, \phi)$, the first two being polar coordinates in the minor cross section and the third the position around the toroidal direction. The conjugate wavenumbers are $(k_r, m, n)$, so that the wave amplitude varies locally as

$$\exp[\mathrm{i}(k_r r + m\theta + n\phi)].$$

In a tokamak the equilibrium is independent of $\phi$, so, as discussed in section 1.2, the corresponding wavenumber $n$ is a constant of the motion. Given $n$, the dispersion relation takes the form

$$D(r, \theta, k_r, m) = 0.$$

If there were another constant of the motion, apart from $D$, even if it could not be identified explicitly, then it would, in principle, enable $m$ to be evaluated in terms of $r$, $\theta$ and $k_r$ and lead to the dispersion relation defining a two-dimensional surface in $(r, \theta, k_r)$ space. If, on the other hand, there is no second constant of the motion, then the ray path can fill a three-dimensional region of this space ergodically. Bonoli and Ott show that either case can occur, depending on the parameters of the tokamak equilibrium, and discuss the implications for accessibility. This behaviour is analogous to the occurrence of chaotic orbits in Hamiltonian dynamical systems. Given the similarity between the equations of ray tracing and Hamilton's equations, it is not surprising that similar phenomena occur.

The approximate dispersion relation (4.5) can be obtained from the full dispersion relation by putting $E_y = 0$. Thus the slow wave is polarised with the electric field predominantly in the $x$–$z$ plane. In order to excite it, in preference to the fast wave, the antenna must be screened to eliminate the $y$ component of the field. A launching system must also be devised which will produce the bulk of the radiated energy in the accessible part of the spectrum. We shall return to the question of antenna design and the theory of the coupling of the antenna to the plasma in a later section, after we have discussed the absorption processes for lower hybrid waves.

## 4.3 Absorption of lower hybrid waves

The cold plasma dispersion relation is adequate to give a description of the way in which the wave travels from the edge of the plasma to the central region where it is to be absorbed, but is not able to describe the absorption process itself, which depends on hot plasma effects. In different regimes lower hybrid waves may deposit energy in either the ions or the electrons, through a variety of mechanisms. In addition, the concentration of energy along resonance cones produces high wave intensities which can lead to non-linear effects being of importance, especially in the low density regions of the plasma. All of this makes lower hybrid heating one of the more complicated of the schemes from the theoretical point of view. The object here will be to introduce the main features, while giving sufficient references to the original literature to enable an interested reader to pursue the details.

The most obvious mechanism of absorption is that arising from the cold plasma lower hybrid resonance, where cold plasma theory would, in fact, predict absorption (Stix 1962). Close to the hybrid resonance $n_x \to \infty$, so that the warm plasma corrections, which depend on $k_x v_{i\,\mathrm{th}}/\Omega_i$, are no longer small. Including these corrections adds terms of higher order in $n_x$ to (4.5) so that, to lowest order in the thermal corrections, the slow wave dispersion relation is changed to something of the form

$$\varepsilon_{\mathrm{th}} n_x^4 + A n_x^2 - B = 0 \qquad (4.7)$$

(Stix 1965), with $\varepsilon_{\mathrm{th}}$ involving contributions from both electrons and ions. Since $\varepsilon_{\mathrm{th}}$ is small, its contribution is only important near the hybrid resonance where $A$ is also small. Its effect is to add a large $n_x$ hot plasma mode which coalesces with the slow mode at the so-called linear turning point, where (4.7) has equal roots, as shown in figure 4.5. Since $\varepsilon_{\mathrm{th}}$ is a small coefficient, this turning point is close to the lower hybrid resonance, defined by $A = 0$.

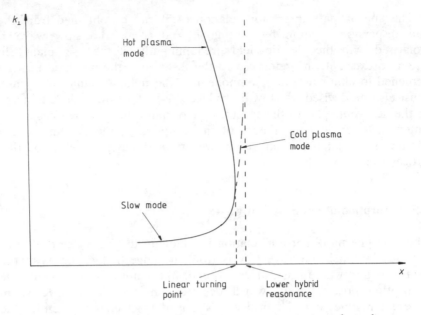

**Figure 4.5** In a hot plasma the slow wave connects to a hot plasma mode before reaching the lower hybrid resonance.

The hot plasma mode has its group velocity in the opposite direction to the slow wave, so that when energy incident on the slow wave approaches the resonance it propagates back towards the lower density region of the plasma on the hot plasma mode. Since $n_x$ is increasing all the time, it was suggested by Stix (1965) that even though the wave was passing through high harmonics of the ion cyclotron heating (typically around $n = 30\text{--}40$) it should be damped by cyclotron damping. It was later shown by Karney and Bers (1977) and Karney (1979) that at the wave amplitudes of interest, non-linear effects destroy the linear cyclotron resonance and that a non-linear process of stochastic ion heating occurs instead.

The essential ideas behind this process can be seen by considering a wave which is propagating perpendicular to the magnetic field and which is purely longitudinal, both of which assumptions are good approximations near the lower hybrid layer. The equations of motion for an ion are then

$$\ddot{x} = \Omega_i \dot{y} + \frac{e}{m_i} E_0 \cos{(kx - \omega t)} \tag{4.8}$$

$$\ddot{y} = -\Omega_i \dot{x}.$$

The second of this pair yields

$$\dot{y} = -\Omega_i x + \text{constant}$$

and a suitable choice of coordinates makes the constant zero, so that the first equation gives

$$\ddot{x} + \Omega_i^2 x = (c/m_i)E_0 \cos(kx - \omega t). \qquad (4.9)$$

There are two limits in which (4.9) leads to periodic particle motion. If $E_0 = 0$, then there is the straightforward cyclotron motion, while if $\Omega_i$ is taken to be zero we may shift to a frame of reference moving with the wave phase velocity, in which the particle is moving in a periodic potential. A particle trapped in one of the potential troughs will then follow a periodic orbit in this frame of reference. As is well known in the theory of dynamical systems (see for example MacKay and Meiss 1987), the non-linear interaction between two periodic motions of this sort may give rise to chaotic behaviour in certain parameter ranges. When this happens, the particle trajectory in phase space wanders through a region of this space in an apparently random manner.

A convenient normalisation of (4.9) is to express time in units of $\Omega_i^{-1}$ and length in terms of $k^{-1}$, so that it becomes

$$\ddot{x} + x = \alpha \cos(x - vt) \qquad (4.10)$$

with $\alpha = E_0 k/(B_0 \Omega_i)$ and $v = \omega/\Omega_i$. If we consider the cyclotron motion alone, then the phase space trajectories are just circles, the radius being the normalised Larmor radius $r$. Interesting dynamical effects occur for particles with $r \geq v$, since at some point during their orbit $\dot{x} = v$ and the driving force on the right-hand side has a point of stationary phase around which a strong resonance may occur. This can be thought of as the particle, as it moves on its cyclotron orbit, having a Landau resonance with the lower hybrid wave, which goes through a large number of complete oscillations during a cyclotron period. The phase space picture of the particle motion is then as shown in figure 4.6, with the particle receiving some increment (positive or negative) to its velocity as it passes through the resonant region.

If $\alpha$ is small, then there is some regular relation between the phases of successive interactions between the particle and the wave, and it follows a regular and predictable path in phase space. However, if $\alpha$ is large enough this phase coherence is lost and the particle follows what looks like a random path through a region of phase space. Even though the particle motion is described by the simple-looking equation (4.9), arbitrarily small differences in its initial conditions can lead eventually to large divergences in the orbits, so that in any practical sense the particle's motion is stochastic.

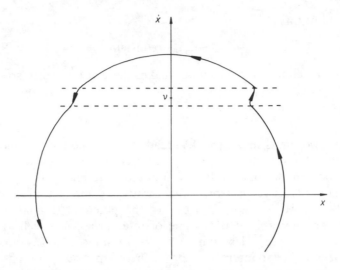

**Figure 4.6** Phase space diagram showing how during ion heating the ion receives velocity increments at positions in its orbit where its velocity is close to the phase velocity of the lower hybrid wave.

The particle interacts strongly with the wave when $|x - v|$ is of the order of $\alpha^{1/2}$, the velocity width within which the particle is trapped in a potential trough of the wave. It is clear from the geometry of figure 4.6 that the particle spends the maximum time in the interaction zone if $r \approx v$, so that this condition gives the minimum threshold for stochastic behaviour. As $r$ increases, the time the particle spends in resonance decreases and the value of $\alpha$ required to produce stochasticity goes up. If $r$ is less than $v - \alpha^{1/2}$ then the particle never enters the resonance region, so this gives a lower bound to the stochastic region. The result is that the boundaries of the stochastic region behave as shown in figure 4.7 (Karney and Bers 1977, Karney 1979). For any value of $\alpha$ above $\alpha_{\min}$ particles within the stochastic region follow orbits in which $r$ wanders in an apparently random fashion. For any given value of $\alpha$ this results in a plateau being produced in the perpendicular distribution function between the values of velocity corresponding to the upper and lower values of $r$ in figure 4.7. Numerical estimates of these bounds are given by Karney (1979).

The region of an inhomogeneous plasma in which this effect occurs can be seen by noting that in physical units the condition

$$r > v$$

becomes

$$k\frac{v_\perp}{\omega} > 1$$

that is, the perpendicular velocity of the particle is greater than the wave phase velocity. The perpendicular phase velocity must therefore be slowed down to a few times the ion thermal velocity before the stochastic heating can affect enough ions to be of any significance. This only happens near the hybrid resonance, when the wave energy goes into the hot plasma mode described above.

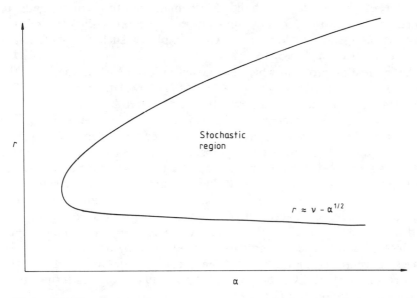

Figure 4.7 The region of parameter space in which stochastic heating of ions takes place.

The other important mechanism for lower hybrid heating is electron Landau damping (Bers 1976). Since the waves propagate into the plasma with $n_\parallel > 1$, the resonance condition $k_\parallel v_\parallel = \omega$ is satisfied for parallel velocities less than $c$ and so the process is possible. The upper limit of the range of resonant velocities is set by the smallest $n_\parallel$ for which the wave can reach the centre of the plasma, that is, by the accessibility condition. The lower limit, on the other hand, is set by the upper end of the $k_\parallel$ spectrum generated by the wave launching system. Generally the range of resonant velocities extends from a few times the thermal velocity up to very high velocities, a substantial fraction of the velocity of light, and Landau damping pulls out a long tail on the parallel distribution function. If the spectrum of $n_\parallel$ is not symmetric about 0, but instead corresponds to waves being launched mainly in one direction around the torus, then this process can be very effective in driving a plasma current. This is probably the aspect of lower hybrid heating which has been of most interest, and is one to which we shall return in the chapter on current drive.

Heating of the bulk of the electrons and of the ions occurs by collisional transfer of energy from the fast electrons. The shape of the electron tail and the rate of heating depends on a balance being set up between the absorption of wave energy by the tail electrons and the transfer of energy to the bulk. Mathematically this corresponds to equating the quasi-linear wave diffusion to the Fokker–Planck collision term as discussed in Chapter 1. An example of such a calculation is given by Fuchs *et al* (1985b).

An interesting problem which has arisen in lower hybrid heating is the so-called spectral gap (Bernabei *et al* 1982, Porkolab *et al* 1984). The problem is that the waves are observed to produce a tail on the distribution, but the spectrum of the waves is such that the lower end of the range of resonant velocities does not appear to be small enough to allow the waves to interact with a substantial number of electrons. Once a tail is established it can be maintained by the lower hybrid waves, but it is not clear how any but a negligible number of particles feel the effect of the waves in the first place. Possible explanations include a shift of $n_{\parallel}$ produced by toroidal geometry and magnetic shear (Colestock and Kulp 1980, Ignat 1981, Bonoli and Englade 1986), non-linear effects produced by the ponderomotive force exerted by the high intensity waves concentrated along the resonance cones (Morales and Lee 1975, Leclert *et al* 1979, Canobbio and Croci 1984) or scattering from fluctuations (Bonoli and Ott 1982).

The first of these is a straightforward consequence of the linear theory of wave propagation. In tokamak geometry the wavenumber parallel to the field is not a conserved quantity, as it is in the slab geometry. The calculations referred to above, making use of numerical solutions of the ray-tracing equations, show that it is possible for this wavenumber component to change quite a lot as the wave goes around the torus.

The ponderomotive force is a non-linear effect. When quadratic terms are included in the equations describing the plasma, the self-interaction of the wave field produces a zero frequency component. This acts as a force, proportional to the square of the wave amplitude, pushing particles towards regions of lower wave intensity. The work of Morales and Lee (1975) looks at the effect of this in producing filamentation. This is a breaking up of the wave front into narrow filaments, resulting from the fact that light is refracted into regions of lower density and that the density perturbation then tends to be enhanced by the effect of the ponderomotive force. If the process is assumed to be slow enough for particles to remain in local thermal equilibrium, the density perturbation is given by

$$\delta n/n = -\tfrac{1}{4}\,\varepsilon_0 |\nabla\phi|^2/(nT)$$

with $T = T_e + T_i$. The waves are assumed to be essentially electrostatic in the frequency range near the lower hybrid frequency and $\phi$ is the potential corresponding to these electrostatic fluctuations. By incorporating this into the density-dependent terms in the wave equation a non-linear equation is obtained. In the work of Morales and Lee this is a modified Kortweg–de Vries equation.

Leclert *et al* take account of the inhomogeneity in the plasma and assume a narrow $k_\parallel$ spectrum so that the wave is quasi-monotonic. A perturbation treatment gives the local dispersion relation to lowest order, and to next order an equation for the slowly varying amplitude of the wave. The non-linear density dependence is introduced at this stage, giving a non-linear equation for the amplitude which can, with a suitable change of variables, be transformed into the non-linear Schrödinger equation. For sufficiently large wave amplitudes, rather larger than occur in practice, soliton solutions are possible. Transforming back to the original spatial variables of the problem reveals that these spread out as they travel into the plasma. The effect of the ponderomotive force in producing a change in the $k_\parallel$ spectrum is discussed specifically by Canobbio and Crocci. The effect of the ponderomotive force on the edge plasma and on the antenna–plasma coupling is discussed by Theilhaber (1982).

There are substantial electron density fluctuations near the plasma edge in a tokamak, scattering from which can be treated as a three-wave interaction problem (Bonoli and Ott 1982), so that the frequency and wavenumber differences between incident and scattered waves equal the fluctuation frequency and wavenumber. The fluctuation wavenumbers are largely perpendicular to the field, so a shift in $k_\perp$ is produced initially. Also, the fluctutions are at a much lower frequency than the lower hybrid waves, so the scattered wave is at essentially the same frequency as the incident wave. If the scattered wave is another slow wave, then, since it has, in effect, the same frequency and parallel wavenumber as the incident wave, it must have the same magnitude of perpendicular wavelength, possibly rotated about the magnetic field direction. The other possibility is coupling to a fast wave, which will lead to a change in the magnitude of the perpendicular wavenumber, as well as its direction. Bonoli and Ott derive the wave kinetic equation describing the non-linear coupling and use it to calculate the probabilities per unit time that the incident wave is scattered either into a slow wave or a fast wave. This information is then coupled into a ray-tracing code, in order to see the effect of scattering in a realistic geometry. The results of this study indicate that scattering from fluctuations could have a very substantial effect, both in terms of allowing parts of the spectrum which would not otherwise satisfy the accessibility condition to reach the centre of the plasma, and of modifying the absorption profile of those rays which can reach the centre of the plasma.

In addition to being scattered from pre-existing fluctuations in the plasma, the lower hybrid radiation may also drive up other modes in the plasma through parametric decay processes (Porkolab 1977). These occur when the incident wave couples to two other modes in the plasma, the sum of whose frequencies and wavenumbers match the frequency and wavenumber of the incident wave. The resonant non-linear interaction amongst the modes then leads to a transfer of energy from the incident pump wave to the other two modes. It is one of the characteristics of such processes that there is a threshold value of the pump wave amplitude below which damping of the decay waves is sufficient to counteract the driving effect of the pump, so that there is no instability. The detailed analysis of Porkolab (1977) shows that the decay processes of most importance involve the interaction between the incident mode, another lower hybrid wave and a low frequency mode of one of two types. These are described by Porkolab as ion cyclotron quasi-modes and non-resonant quasi-modes. The ion cyclotron quasi-modes are electrostatic ion cyclotron waves, which are very strongly damped by both Landau and cyclotron damping, while the non-resonant modes are not properly modes of the plasma and only exist in the presence of the pump wave. They can be driven up by the pump wave, and by absorbing energy and momentum allow the parametric decay process to take place. For these modes the other decay wave may not only be a cold plasma lower hybrid wave, but also the hot ion branch, discussed earlier in this section, or an ion Bernstein mode.

Porkolab derives the non-linear dispersion relation describing the parametric decay process and the thresholds for the various possible regimes. A realistic estimate of the latter requires that the finite extent of the pump wave, which, we should remember, is confined within the resonance cones, be taken into account. This is because the decay waves convect energy out of the spatial region occupied by the pump, leading to a threshold which increases as the extent of the pump becomes smaller. With all these effects taken into account, it is found that the process with the lowest threshold, likely to present the greatest danger to lower hybrid heating schemes, is the decay to non-resonant quasi-modes and hot ion lower hybrid waves (see also Villalon and Bers 1980). This occurs if the incident frequency is less than twice the maximum value of the lower hybrid frequency in the plasma. (Remember that it is not essential for lower hybrid heating that the incident frequency be equal to the lower hybrid frequency somewhere in the plasma. Electron Landau damping does not depend on proximity to the cold plasma resonance.) Since the lower hybrid frequency increases with density, this implies the existence of a density limit with a given incident frequency, above which parametric decay becomes a significant loss process. The energy of the incoming wave goes mainly to the hot ion modes, which are the same as those produced by the linear mode-conversion process, and are absorbed in a similar way. However, since the parametric decay takes place before the lower hybrid layer is reached, the result may be that energy is deposited nearer the plasma edge than desired. If the parameters of the system are such that the heating would otherwise be through electron Landau damping, parametric decays can change it to ion heating. This has important implications for current drive schemes and we shall return to it in Chapter 6.

As we have seen, absorption of lower hybrid waves can take place on either electrons or ions, and the conditions under which one or other of these absorption processes is dominant are particularly important if an asymmetric wave spectrum is being used with a view to producing current drive as well as heating. Electron Landau damping produces a long tail on the parallel distribution function which leads to very effective current drive. On the other hand, ion damping produces an increase in the perpendicular ion velocities, with no effect on the current.

The switch between the two absorption mechanisms has been analysed by Wegrowe and Tonon (1983), who show that the ratio of $\omega/(k_\parallel v_{\mathrm{th\,e}})$ to $\omega/(k_\perp v_{\mathrm{th\,i}})$ increases with density. The first of these factors determines the rate of electron Landau damping and the second the rate of ion stochastic heating. Wegrowe and Tonon suggest that the switch between electron and ion heating takes place when the two parameters are of the same order, and show that this criterion gives reasonable agreement with the experimentally observed critical density above which lower hybrid current drive is quenched. However, the discussion given above shows that parametric instabilities also become more significant with increasing density. The criterion that these occur strongly is very similar to the criterion of Wegrowe and Tonon, so that they may also play a role in quenching the current drive.

## 4.4 Launching of lower hybrid waves

The accessibility properties of lower hybrid waves, as discussed in section 4.1, require that a spectrum be launched with a range of $n_\parallel$ greater than one, and with the waves polarised so that the electric field is parallel to the magnetic field of the plasma. The system used to generate such waves is the so-called grill, first suggested by Lallia

(1974). This consists of a phased array of rectangular waveguides with their long sides along the poloidal direction, rather as shown in figure 4.8. An analysis of this structure was given by Brambilla (1976, 1979) who calculated the spectral density of the power flux radiated by such an antenna. Such a calculation is a vital component of any analysis of lower hybrid heating and current drive in a tokamak, since a knowledge of the spectrum produced by the antenna is esssential for calculation of the electric field patterns within the plasma and the energy absorption profile. In what follows we give an outline of Brambilla's calculation. A more detailed three-dimensional analysis of the waveguide coupling problem is given by Bers and Theilhaber (1983).

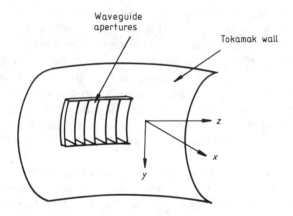

**Figure 4.8** Schematic diagram of a grill for launching lower hybrid waves, together with the coordinate system used in the theoretical analysis of it. The toroidal magnetic field is in the $z$ direction.

Cartesian coordinates are taken as shown in figure 4.8, with the simplifying assumption that the waveguides are infinite in the $y$ direction and that the curvature of the tokamak wall can be neglected, so that the apertures of the array lie in the $y$–$z$ plane. The fields in the guide are represented as a sum of waveguide eigenmodes of the form

$$E_z = \sum_{p=1}^{N} \theta_p(z) \sum_{n=0}^{\infty} (\alpha_{np}^{TM} e^{-\gamma_n x} + \beta_{np}^{TM} e^{\gamma_n x}) \cos[n\pi(z - z_p)/b]$$

$$E_y = \sum_{p=1}^{N} \theta_p(z) \sum_{n=0}^{\infty} (\alpha_{np}^{TE} e^{-\gamma_n x} + \beta_{np}^{TE} e^{\gamma_n x}) \sin[n\pi(z - z_p)/b] \quad (4.11)$$

with similar expressions for the magnetic field components. In (4.11) $p$ numbers the waveguides and $b$ is the width of each aperture in the toroidal direction. The function $\theta_p$ is one in the range $|z - z_p| < b/2$

occupied by the $p$th guide and zero elsewhere. The sum over $n$ corresponds to a sum over the eigenmodes of an individual guide. The factor $\gamma_n$ is given by

$$\gamma_n^2 = \frac{n^2\pi^2}{b^2} - k_0^2 \qquad (4.12)$$

$(k_0 - \omega/c)$ so that the waves propagate for $n\pi < k_0 b$. For $n\pi < k_0 b$, it is assumed that the coefficients $\alpha_{np}$ are known and represent the mode amplitudes propagating towards the plasma. For $n\pi > k_0 b$, $\alpha_{np} = 0$. The object of the theory is to evaluate the coefficients $\beta_{np}$, which represent the reflection coefficients of any propagating modes and the level of excitation of evanescent modes with $n\pi > k_0 b$.

Between the wall and the plasma it is assumed that there is a vacuum region, within which the fields may be written

$$E_z = \int_{-\infty}^{\infty} \mathrm{d}k_\parallel e^{ik_\parallel z}[E_z(k_\parallel)(e^{ik_\perp x} + \rho_{11}(k_\parallel)e^{-ik_\perp x}) + E_y(k_\parallel)\rho_{12}e^{-ik_\perp x}]$$

$$E_y = \int_{-\infty}^{\infty} \mathrm{d}k_\parallel e^{ik_\parallel z}[E_z(k_\parallel)\rho_{21}e^{-ik_\perp x} + E_y(k_\parallel)(e^{ik_\perp x} + \rho_{22}e^{-ik_\perp x})]$$

$$(4.13)$$

Here $k_\perp^2 = k_0^2 - k_\parallel^2$ and the $\rho_{ij}$ are the elements of the reflection matrix for plane waves from the plasma surface. Note that we allow for the fact that the plasma couples the $y$ and $z$ components of the electric field, so that the reflected part of the $z$ component contains a contribution from the incident $y$ component and vice versa.

At the tokamak wall the expressions (4.11) and (4.13) must coincide. This, together with the continuity of $B_y$ and $B_z$ at the waveguide mouth gives a set of linear algebraic equations for the $\beta$'s in (4.11). After truncation of this infinite system to include only a finite number of evanescent modes, the system can be solved numerically to obtain the coupling properties of the grill. This, of course, assumes that the reflection coefficients $\rho_{ij}$ are known, which requires an analysis of the propagation of the waves into the plasma. It is assumed that any power which propagates towards the centre of the plasma is either absorbed, or deflected away from the antenna, so that the only reflection is a result of the narrow evanescent region at the plasma edge. Beyond this region, where both wave modes propagate, the boundary condition to be imposed is that the waves travel away from the antenna.

Since an explicit treatment of the wave absorption is not required, cold plasma theory can be used and the behaviour of the wave fields described by a pair of coupled second-order equations from $E_y$ and $E_z$ derived in a straightforward way from Maxwell's equations and the cold plasma conductivity tensor. At low densities the coupling is weak, and to a good approximation there are separate equations for $E_y$, giving the fast wave, and $E_z$, giving the slow wave (Brambilla 1979). An analytic approximation for the slow wave can be found if the density gradient around the slow wave cut-off is assumed to be linear, since the wave equation then has solutions in terms of Airy functions. The fast wave cut-off is further into the plasma, but again a solution in terms of Airy functions can be found if the rapid change in the fast wave refractive index near the cut-off is approximated by a linear function. Given the experimental uncertainties in the density profile near the edge, it is pointed out by Brambilla that such simplifications will give quite adequate results.

This theory gives a good description of the behaviour of lower hybrid antennae and enables them to be designed in such a way as to optimise the coupling. Excitation of waves in the inaccessible region of the spectrum is to be avoided as much as possible since they produce surface modes localised between the plasma edge and the point of confluence of the cold plasma waves. This can be done by increasing the number of waveguides in the grill, but on the other hand, this produces more energy in high $n_\parallel$ modes which are more strongly reflected from the plasma edge. If current drive is the object of the system, a spectrum asymmetrical in the parallel wavenumber can be produced by suitable phasing of the waveguides which make up the antenna.

## 4.5 Experimental results on lower hybrid heating

Some of the basic properties of lower hybrid wave propagation, in particular the formation of a well collimated beam determined by the resonance cones, were verified in a series of experiments by Bellan and Porkolab (1974, 1976). These experiments were carried out in a relatively cold cylindrical plasma column, with conducting rings surrounding the plasma serving as antennae to generate the slow wave. Eight such rings were used, with alternate ones excited 180° out of phase and the separation chosen to give a suitable range of parallel wavenumbers. The ring assembly was short enough to allow a reasonable space for the waves to propagate beyond its ends and exhibit the resonance cone structure sketched in figure 4.3. A number of probes were used to measure the wave fields at various positions, and showed quite clearly that the resonance cones occurred and that their shape was in agreement with theoretical predictions. The damping of the waves was also measured and was found to be consistent with an electron distribution function in which around 30% of the electrons were pulled out into a tail with a parallel temperature about six times the bulk temperature.

Independent evidence for such a distribution was seen from measurements made with a Langmuir probe and with a parallel electron energy analyser. The wave was damped before it reached the lower hybrid layer, so that it was impossible to investigate the conversion of the slow wave into a hot plasma wave.

At about the same time experiments on a tokamak (Bernabei *et al* 1975) verified Brambilla's theory of waveguide coupling and showed that efficient coupling of the wave energy into the plasma could be achieved. Lower hybrid heating is in a frequency range where high power continuous wave sources are readily available, and has the further advantage that the launching structure may be mounted flush with the wall, or only slightly out from it, but still within the shadow of the limiter. This minimises the problem of impurities produced by the interaction between the plasma and the antenna structure, and helps make lower hybrid heating a technologically attractive scheme. As a result, these early experiments have been followed by a substantial number of experiments at higher powers, and an excellent review of progress up to about 1984 is given by Porkolab (1985b).

Grill systems consisting of two to eight waveguides were generally used in early experiments, but more recent machines may have many more (see the comment below about JT60). Some sort of bellows arrangement is usually incorporated in order that radial adjustment of the position of the waveguide apertures is possible. By altering the gap between the plasma and the antenna by means of the bellows, it is possible to match the plasma impedance to the antenna impedence so as to minimise reflection. The waveguides are excited in such a way that the electric field is in the toroidal direction, the narrow dimension of the apertures, and the phase varies between adjacent waveguides in a controlled way. This can be done by driving all the amplifiers in the system with a common oscillator, introducing phase shifts where necessary. So, if the adjacent guides are excited out of phase by $\pi$ a standing wave pattern is set up in the near-field zone and the wave is launched symmetrically with respect to the two directions around the torus. On the other hand a $\pi/2$ phase shift produces a travelling wave pattern and results in waves being launched mainly in one direction. With a properly matched antenna, excellent coupling to the plasma is obtained with less than 20% of the incident power being reflected. Recently the large Japanese JT60 tokamak has been constructed with launching systems consisting of a much larger number of waveguides than the earlier experiments. This produces a very narrow spectrum, which can be localised around a value of $n_{\parallel}$ for which there are good accessibility properties.

As we have indicated in the discussion of the theory of lower hybrid waves, we may expect them to interact with either electrons or ions, depending on the parameters of the experiment. Experiments aimed at ion heating have proved somewhat unpredictable in the results they give (Porkolab 1985b), ranging from some which give excellent results to others in which the heating is almost non-existent. This is likely to be a result of the fact that the regime in which ion heating is predicted is also that in which parametric instabilities and scattering of waves by fluctuations in the edge region of the plasma are also important. These effects could prevent energy from reaching the central part of the plasma. Since they depend on the details of the plasma in the edge region, the results might be expected to be sensitive to the precise nature of the plasma and to vary from one machine to another.

Electron Landau damping takes place if the parallel phase velocity of the wave is less than around three times the thermal velocity. In order that the wave be absorbed in the centre of the plasma, it is necessary that $n_\parallel$ is not too large. The parallel phase velocity is inversely proportional to $n_\parallel$, so that the larger the latter is, the lower the temperature at which strong Landau damping can occur. To avoid strong damping in the relatively cool plasma near the wall, and to confine Landau damping to the hot plasma in the central region of the tokamak, it is necessary to place an upper limit on the range of $n_\parallel$ in the spectrum. On the other hand, the accessibility condition imposes a minimum value of $n_\parallel$. For current tokamaks there is a comfortable gap between these upper and lower values and the electron heating regime is easily accessible. However, as the temperature goes up towards the reactor regime, the window of acceptable values of $n_\parallel$ becomes rather small, and it is unlikely that much energy will get beyond the layer at which the temperature reaches 10 keV. The temperature in the centre of a reactor will be greater than this, and for this reason there is interest in using the fast wave rather than the slow wave for heating and current drive in this regime. We shall discuss the subject of the fast wave in the next section.

For present-day tokamaks, the antenna can easily be arranged to give a spectrum in the correct range and ion heating and parametric instabilities can be avoided by taking $\omega > 2\omega_{LH}$ and having the electron temperature greater than around one third of the ion temperature. Results in various tokamaks, with megawatt power levels, have shown that in this regime very effective heating of the bulk electrons near the plasma centre can be produced, accompanied by significant ion heating as a result of collisional equilibration. The bulk electron temperature rise may be monitored in various ways, using x-ray emission, Thomson scattering and electron cyclotron absorption measurements. The ion heating can be detected by neutron emission and charge exchange analysis. Typically the effect of radiofrequency heating is distinguished from that of Ohmic heating by switching on the heating for a short period during a longer Ohmic discharge. During the radiofrequency discharge there is an increase in the temperature which decays after the pulse is switched off, at a rate determined by the energy confinement time of the discharge.

In order to produce a temperature rise in the core of the plasma in a predictable way, it appears that the electron heating rather than the ion heating mode should be chosen. Before the latter can be used reliably it will be necessary to understand in greater detail the conditions under which edge interactions of various sorts inhibit the propagation of the wave to the lower hybrid layer.

## 4.6 Fast wave heating

As already mentioned in the last section, the condition for strong Landau damping, $\omega/k_\parallel < 3v_{\mathrm{th\,e}}$, means that if the wave is not to be heavily damped at the plasma edge, $n_\parallel$ must be small enough that this condition is not satisfied until the centre of the plasma is reached. On the other hand, accessibility of the plasma core to the slow wave imposes a minimum value of $n_\parallel$, and for reactor temperature plasmas there is a very narrow window of values of $n_\parallel$ for which the desired behaviour occurs. Above about 10 keV, this window essentially disappears. For this reason use of the fast wave rather than the slow wave has been suggested (Theilhaber and Bers 1980). Another reason to investigate the fast wave is that the slow wave frequency must be at least the lower hybrid frequency at the centre of the plasma, since the wave only propagates above the lower hybrid frequency. For high-density, high-magnetic-field reactor designs this implies a rather high frequency and hence rather small waveguide apertures. Since the fast wave propagates through the lower hybrid layer, it may be used at lower frequencies. The available range of frequencies extends upwards from a few times the ion cyclotron frequency, the desired mode being simply the extension to higher frequencies of that used for ion cyclotron heating.

Referring back to figure 4.1, it can be seen that for the wave to propagate to the centre of the plasma we must still be in the regime of figure 4.1(c), so that the same accessibility limit on $n_\parallel$ is relevant. However, the lower the frequency, the closer this critical value becomes to unity, so by being able to work at frequencies below the lower hybrid, it may be possible to extend the useful range of $n_\parallel$ downwards. In order to excite the fast rather then the slow wave, the electric field must be perpendicular to the magnetic field, that is, in the poloidal direction rather than the toroidal direction as was the case for the slow wave. This makes the launching of the wave more difficult than for the slow wave, at least at the higher end of the frequency range. To have a reasonably uniform electric field across the aperture of the waveguide, using a rectangular guide as in the slow wave grill, the electric field must be parallel to the short side, since the parallel component of the field is small at the wall. So the required polarization needs the long side to be in the toroidal direction. However, in a vacuum guide the long dimension is greater than half a wavelength, in order that the wave may propagate, and this conflicts with the requirement that we generate a spectrum with the bulk of the energy in $n_\parallel^2 > 1$. Waves with $n_\parallel^2 < 1$ propagate in the vacuum, but have a cut-off near the edge of the plasma. They can undergo multiple reflections between the cut-off and the edge and lead to a large reactive load on the source.

To generate waves with the required spectrum and polarization is thus more complicated than simply turning the lower hybrid grill through 90°. One answer to the problem of generating the required $k_\parallel$ spectrum is to use dielectric-filled waveguides. The reduction in wave phase velocity in the dielectric means that the dimensions of the waveguide can be reduced in order to correspond to a $n_\parallel$ greater than one (Theilhaber and Bers 1980). Such a launching system has been used on the PLT machine at Princeton (Pinsker *et al* 1987). Other methods which have been used involve loop antennae (Pinsker *et al* 1987) and slotted waveguides (Colborn *et al* 1987). In the former scheme the wave is launched by means of a series of metallic loops, part of whose length is exposed to the plasma. The current in the loops flows in the toroidal direction, this being the appropriate orientation for coupling to the magnetic field of the fast wave. A series of these (six in the experiment referred to above) can be excited with phase differences giving the required parallel spectrum, just like the slow wave grill. The slotted waveguide consists of a rectangular waveguide running around part of the circumference of the tokamak in the toroidal direction with its broad side facing the plasma. Longitudinal slots cut in this face can, if the dimensions and shape are correctly chosen, radiate energy with the required spectral characteristics towards the plasma.

Another problem in getting the fast wave to couple to the plasma can be seen by referring again to figure 4.1, which shows that the cut-off, beyond which the wave propagates, is further from the plasma edge than for the slow wave. The wave must tunnel through this evanescent region before it can reach the interior of the plasma. As $n_\parallel$ increases the tunnelling becomes less, a result which might be anticipated from the fact that in a vacuum the amplitude decays as

$$\exp\left[-\left(n_\parallel^2 - 1\right)^{1/2} \omega/c\right].$$

Theilhaber and Bers (1980) analyse the tunnelling with the assumption of a linear density gradient. The idea of the calculation is similar to that for slow wave coupling which we have already discussed. The linear density gradient gives rise to a parabolic variation of the fast wave refractive index near the plasma edge, and an analytic solution can be found in terms of parabolic cylinder functions. From this an estimate can be made of the upper limit of the $n_\parallel$ spectrum for which there is satisfactory coupling. The result is

$$|n_\parallel| < \beta^{1/3}$$

where

$$\beta = \frac{c}{\omega}\,\frac{\mathrm{d}}{\mathrm{d}x}\left(\frac{\omega_{pe}^2}{\omega\Omega_e}\right)$$

evaluated at the plasma edge. Numerical calculations of antenna–plasma coupling for both fast and slow waves have been carried out by Pinsker *et al* (1986).

The fast wave may be damped by Landau damping. This is weaker than for the slow wave, since the polarization of the wave is such that the parallel electric field component is much smaller than the perpendicular component. However for hot reactor-type plasmas it should be sufficient to produce effective absorption of the wave. Fast waves do not form resonance cones and the energy spreads out from the antennae, so that non-linear effects are less likely to occur. As in the case of the slow wave, there is a lower limit to the values of $n_\parallel$ for which the central region of the plasma is accessible. However, because the fast wave may be used at a lower frequency, the accessibility limit can be made close to one. This, in turn, raises the resonant velocities for Landau damping and allows the wave to penetrate to regions of higher temperature.

Several experiments have been carried out to look at fast wave current drive (Ohkubo *et al* 1986, Pinsker *et al* 1987, Colborn *et al* 1987, Uesugi *et al* 1987). In general, the results seem to show a density limit for current drive of the same order as that expected for the slow wave, and at the time of writing there seems to be no unequivocal evidence showing excitation of a fast wave which gives heating and current drive with properties notably different from the slow wave. This is a problem on which further work may be expected in the near future. The perceived advantages of the fast wave in very high temperature plasmas, and especially the prospect of using it to produce current drive in a regime where the slow wave cannot penetrate to the centre of the plasma, mean that it is of great interest to designers of tokamak reactors and of the next generation of machines in which it is planned that conditions close to those of a reactor should be reached.

The position as regards lower hybrid heating may be summed up by saying that it works well in the lower density regime in which absorption is by Landau damping. In the same regime it has also been very successful in driving current, as we shall see in Chapter 6. At higher densities, when ion heating is expected, edge effects and parametric instabilities seem to produce results which are rather erratic and suggest that this may not be a reliable way to heat the plasma. The fact that the slow wave is predicted to be unable to penetrate a reactor-temperature plasma has led to a great deal of interest in the scheme using the fast wave instead of the slow wave, though at present its theoretical advantages do not seem to have been demonstrated experimentally. This is an area in which a substantial research effort may be expected in the future.

# 5  Electron Cyclotron Heating

## 5.1 Introduction

With electron cyclotron heating we arrive at the highest range of frequencies used for plasma heating. The frequency required is generally the fundamental or the second harmonic of the electron cyclotron frequency, which for typical tokamak magnetic fields is in the range 30–150 GHz. This very high frequency has proved something of a hindrance to the development of this heating scheme. In the lower frequency ranges the technology required to produce high power radiation was already in existence, these being frequencies appropriate to radar, microwave communications and similar applications. High power sources in the electron cyclotron range of frequencies are of more recent development, and it is only within the past few years that sources giving hundreds of kilowatts at frequencies of tens of gigahertz have become readily available. These are devices called gyrotrons, which operate by extracting energy from a beam of relativistic electrons in a strong magnetic field. An account of the development of these systems and further references are given by Kreischer *et al* (1985). Another possible source is a free electron laser, which can produce very high powers at frequencies in excess of 100 GHz. An experiment in which such a device will be used is described by Thomassen (1988).

Although wave generation at the lower frequencies which we have discussed is not a problem, our discussion has demonstrated that coupling of the wave to the plasma can be a problem and that careful control of the plasma and antenna properties is necessary to avoid problems like plasma heating at the edge and the resulting increase of the impurity flux from the wall. Given the existence of suitable sources, however, electron cyclotron heating has a number of advantages in this respect. The problems at lower frequencies are basically a consequence of the fact that the wave is evanescent in vacuum, and has to tunnel through a region in which it is non-propagating in order to reach the plasma interior. If the layer in which the wave is evanescent is too large, then a significant fraction of the incident power may be reflected, and a large amplitude standing wave set up near the edge. At the frequencies involved in electron cyclotron heating, and the short wavelengths which accompany them, tunnelling is impossible, except for very small devices. The wave must, therefore, be such as to propagate in vacuum and at the plasma edge. No sophisticated coupling scheme is required to get it into the plasma, and it can be launched from a simple waveguide aperture. This has the advantage of avoiding internal structures which may contribute to the problem of impurities. Another possible advantage of electron cyclotron heating arises from the fact that the absorption is quite narrowly localised around the cyclotron resonance layer. Such localised heating may be of use in controlling the current profile in a tokamak, with a view to suppressing magnetohydrodynamic instabilities.

So far as absorption is concerned, the process of cyclotron resonance is similar to the ion cyclotron case. Electrons interact strongly with a right circularly polarised field component, in the opposite sense to that which interacts with the ions. As was the case for ions, the circularly polarised component which interacts strongly with the particles is shorted out in a cold plasma, and absorption is the result of thermal effects. The main difference is that relativistic effects are important. This is because the interaction between waves and particles depends on a matching of the wave frequency and the particle gyrofrequency. The latter is changed by the relativistic velocity dependence of the electron mass, and a very small change may be significant, in the sense of being comparable with the Doppler shift over a significant range of values of $n_\parallel$. The result is that relativistic effects are important, even at temperatures at which the thermal velocity would not normally be thought of as being in the relativistic regime. The properties of wave propagation and absorption will be described in some detail in the following sections.

## 5.2 Cold plasma propagation

Although absorption is dependent on hot plasma effects, the propagation of the waves up to the cyclotron resonance is essentially determined by the cold plasma dispersion relation. We shall, therefore, look first at cold plasma propagation in order to consider such basic questions as in what parameter ranges the resonance is actually accessible to a wave. In a homogeneous plasma the two possible wave modes are given by the well known Appleton–Hartree dispersion relation (Stix 1962, Allis *et al* 1963) for a wave propagating at an angle $\theta$ to the steady magnetic field

$$n^2 = \frac{k^2 c^2}{\omega^2} = 1 - \frac{2\alpha\omega^2(1-\alpha)}{2\omega^2(1-\alpha) - \Omega_e^2 \sin^2\theta \pm \Omega_e \Gamma} \tag{5.1}$$

where

$$\Gamma = (\Omega_e^2 \sin^4\theta + 4\omega^2(1-\alpha)^2 \cos^2\theta)^{1/2}$$

and

$$\alpha = \frac{\omega_{pe}^2}{\omega^2}.$$

The + sign in (5.1) gives the ordinary mode, which is independent of the magnitude of the magnetic field for $\theta = \pi/2$, and the − sign the extraordinary mode. The essential features of electron cyclotron propagation can be seen most easily by specialising to the case of perpendicular propagation, in which (5.1) simplifies to

$$n^2 = 1 - \frac{\omega_p^2}{\omega^2} \tag{5.2}$$

for the ordinary mode and

$$n^2 = \frac{(\omega_p^2 - \omega^2)^2 - \Omega_e^2 \omega^2}{\omega^2(\omega^2 - \omega_p^2 - \Omega_e^2)} \tag{5.3}$$

for the extraordinary mode.

The behaviour of the O mode is simply that the wave will propagate until the density reaches the point where $\omega_p^2 = \omega^2$. If we wish to have absorption at the fundamental then there is a density limit defined by the condition

$$\omega_p^2 < \Omega_e^2 \tag{5.4}$$

and for the second harmonic

$$\omega_p^2 < 4\Omega_e^2. \tag{5.5}$$

For tokamaks, the plasma and electron cyclotron frequencies in the centre of the plasma are generally of the same order of magnitude. This means that these density limits are within the normal operating range and so must be taken into consideration in electron cyclotron heating schemes.

For the X mode the behaviour is slightly more complicated since now there is a cut-off where

$$(\omega_p^2 - \omega^2)^2 = \Omega_e^2 \omega^2 \tag{5.6}$$

and a resonance, the upper hybrid resonance, where

$$\omega^2 = \omega_{pe}^2 + \Omega_e^2. \tag{5.7}$$

At non-zero $n_\parallel$ the dispersion relation can be written in the form of equation (3.1) and it can be seen that the O-mode cut-off and upper hybrid resonance are unaffected. However, the condition for the cut-off of the X mode is changed to

$$\left(1 - \frac{\omega_{pe}^2}{\omega^2 - \Omega_e^2} - n_{\parallel}^2\right)^2 - \left(\frac{\omega_{pe}^2\Omega_e}{\omega(\omega^2 - \Omega_e^2)}\right)^2 = 0$$

which gives

$$[\omega^2(1 - n_{\parallel}^2) - \omega_p^2]^2 = \omega^2\Omega_e^2(1 - n_{\parallel}^2)^2. \qquad (5.8)$$

To see how the cut-offs and resonance affect propagation across an inhomogeneous magnetic field we draw the Clemmow–Mullally–Allis (CMA) diagram, figure 5.1, in which these are plotted in terms of $X = \omega_p^2/\omega^2$, $Y = \Omega_e^2/\omega^2$; effectively in terms of the density and magnetic field. The utility of such a diagram is that we can look at the path on it which must be followed by a ray travelling from the plasma edge to the cyclotron resonance and see whether it can so travel without encountering a cut-off or a resonance.

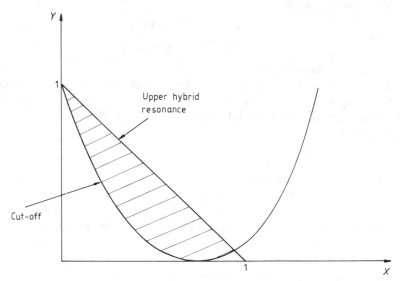

**Figure 5.1** CMA diagram for the X mode. The wave does not propagate in the shaded region or in the region to the right of all the curves.

The extraordinary mode does not propagate in the shaded region. Except in a very small machine, accessibility for the X mode requires that the path traced out on this diagram by the parameters $X$ and $Y$ between the edge of the plasma and the desired resonance must not cross this region. At the edge of the plasma $X = 0$, since the density goes to zero, so we always start on the $Y$ axis. The magnetic field increases from the outside to the inside of a tokamak, so if the wave is launched from the outside the path goes upwards, while with inside launching it moves downwards. At the fundamental $Y = 1$, and it is clear that this is only accessible from the high field side. There is a density limit, since the ray must reach the cyclotron resonance before it reaches the cut-off on the high density side. Using $n_\parallel = 0$ to estimate this density limit gives the condition that $X$ has to be less than two when $Y = 1$, i.e.

$$\frac{\omega_p^2}{\Omega_e^2} < 2.$$

The other scheme in common use is the second harmonic, corresponding to $Y = \frac{1}{4}$. This is accessible from both the outside and inside. From the outside, the resonance must be reached before the left-hand branch of the cut-off is reached, giving the condition $X < \frac{1}{2}$ when $\omega = 2\Omega_e$, i.e.

$$\omega_p^2 < \frac{\omega^2}{2} = 2\Omega_e^2.$$

From the inside we require $X < \frac{3}{2}$ when $\omega = 2\Omega_e$ so that

$$\omega_p^2 < \frac{3}{2}\omega^2 = 6\Omega_e^2.$$

Inside launch of the X mode has the advantage that at both the fundamental and second harmonic it can reach higher densities than the O mode, which has the same density limit regardless of the direction of propagation. On the other hand, inside launch is technically less convenient. For a given magnetic field there are advantages in going to the second harmonic, since this raises the density limit. To set against this, however, there are the problems of obtaining a source at the higher frequency.

At non-zero $k_\parallel$ it is possible for the wavenumbers of the O and X modes to coincide at some point in the plasma and for mode conversion to take place from one to the other (Preinhalter and Kopecky 1973, Weitzner and Batchelor 1979). This allows an ordinary mode to propagate to the interior of the plasma, then convert to an extraordinary mode, which is absorbed at the upper hybrid resonance. The mechanism can be explained as follows, following essentially the treatment of Weitzner and Batchelor (1979). The cut-off condition $\omega^2 = \omega_p^2$ for the O mode and condition (5.8) are satisfied at the same point in the plasma if

$$(1 - n_\parallel^2)(1 + \Omega_e^2/\omega_p^2) = 1$$

when $\omega^2 = \omega_p^2$. The wave dispersion curves at this cut-off then look as shown in figure 5.2. A wave incident in the O mode is partially reflected in the X mode which, having its group and phase velocities in opposite directions near the cut-off, continues to propagate in the same direction as the incident wave, but with its phase velocity in the opposite direction. When the density increases to a value at which $\omega/\omega_p$ is equal to the value at the minimum on the X mode branch of figure 5.2, the group velocity reverses and the wave travels back towards the low density region, eventually reaching the position of the upper hybrid resonance. At this point it might be expected to turn into a hot plasma Bernstein mode and eventually be absorbed. Weitzner and Batchelor consider the efficiency of this process, coming to the conclusion that, in reactor conditions, it is only possible for a very narrow band of $k_\parallel$, and so is unlikely to be of practical interest. We can therefore confine our interest to the case of simple propagation of one of the cold plasma modes up to the point where its frequency is a multiple of the cyclotron frequency and where hot plasma effects become important in determining how it is absorbed.

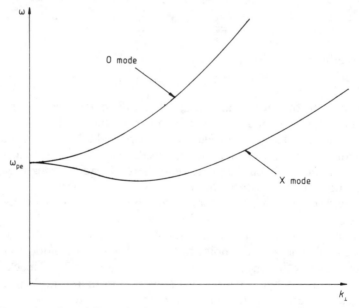

**Figure 5.2** O and X mode dispersion curves for the value of parallel refractive index at which mode conversion between the two is possible.

## 5.3 Electron cyclotron wave propagation in a hot plasma

The main complication here, as compared with the lower frequency regimes, is that in order to obtain the correct absorption profiles it is necessary to take account of the relativistic mass dependence of the electrons, and to use relativistic forms of the dielectric tensor elements. We shall begin with a simple discussion, aimed at explaining the physical effects associated with the relativistic mass dependence, then discuss in more detail the form taken by the dielectric tensor elements, since these are not so familiar as the non-relativistic version quoted in Chapter 1. For most realistic tokamak parameters, the thermal velocity is much less than the velocity of light, allowing various approximations valid in weakly relativistic plasmas to be used.

If we consider radiation near the $n$th harmonic of the cyclotron frequency, then damping occurs in the non-relativistic theory as the result of the imaginary part in resonant integrals of the form

$$\int \frac{F(v)}{\omega - k_\| v_\| - n\Omega} \, d^3v \qquad (5.9)$$

or, physically, as a resonance between the wave and those particles whose parallel velocity is such that the doppler-shifted wave frequency which they see is a multiple of the cyclotron frequency. As $k_\| \to 0$, all particles interact strongly and the integral (5.9) becomes real, but proportional to $1/(\omega - n\Omega)$. The result is that close to the harmonics the dispersion curves diverge from the cold plasma behaviour and connect to the Bernstein modes, which are undamped for perpendicular propagation but have resonances (i.e. their wavenumber tends to infinity) at the cyclotron harmonics.

If we now take account of the relativistic mass dependence, and let $\Omega_0$ be the cyclotron frequency calculated using the electron rest mass, the resonance condition becomes

$$\omega - k_\| v_\| - n\Omega_0 \left(1 - \frac{v^2}{c^2}\right)^{1/2} = 0$$

or for weakly relativistic particles

$$\omega - k_\| v_\| - n\Omega_0 \left(1 - \frac{1}{2}\frac{v^2}{c^2}\right) = 0. \qquad (5.10)$$

The result is that even at perpendicular incidence the resonant integrals corresponding to (5.9) contain velocity-dependent denominators and must be treated in the same way as the familiar Landau integrals.

An approximate indication of the importance of the relativistic effect can be given simply by comparing the Doppler and relativistic frequency shifts in (5.10). If they are approximately equal, for typical particle velocities, then

$$k_\| v_{th} \approx \frac{1}{2} n\Omega \frac{v_{th}^2}{c^2} \approx \frac{1}{2} \frac{\omega v_{th}^2}{c^2}$$

or

$$n_\parallel \approx \frac{v_{th}}{c} \approx \left(\frac{T_e(\text{keV})}{500}\right)^{1/2}. \tag{5.11}$$

For temperatures of a few keV it can be seen that (5.11) implies that over a significant cone of angles around the perpendicular direction the relativistic effects are important. (For small angles, $n_\parallel$ is approximately the angle, in radians, between the launch direction of the waves and the perpendicular.) This simple argument tends to underestimate the importance of relativistic effects, and a somewhat more detailed discussion of their importance is given later.

Some important qualitative results about the nature of the absorption profile can be obtained by solving (5.10) for $v_\parallel$ to give

$$\frac{v_\parallel}{c} = n_\parallel \pm \left(n_\parallel^2 - \frac{v_\perp^2}{c^2} + \frac{2(\Omega_0 - \omega)}{\Omega_0}\right)^{1/2}. \tag{5.12}$$

If we plot the resonant values of $v$ as a function of the magnetic field strength in the vicinity of the point $(\Omega_0 - \omega)/\Omega_0 = 0$ we obtain a series of nested parabolae, corresponding to different values of $v_\perp$ with that for $v_\perp = 0$ on the outside (Cairns *et al* 1983), as illustrated in figure 5.3.

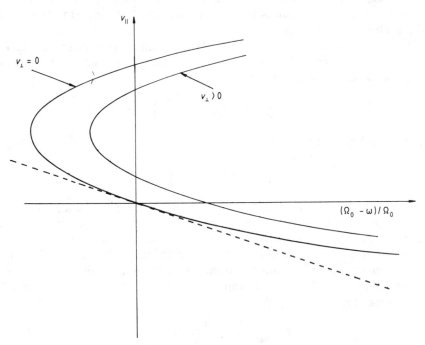

**Figure 5.3** Values of the parallel velocity satisfying the relativistic resonance condition. The broken line shows the non-relativistic result.

The non-relativistic equivalent, which is independent of $v_\perp$, is given by the dotted line tangent to the outer parabola at the origin. For this to be a good approximation we require not only $v^2/c^2 \ll n_\parallel^2$ but, as can be seen from (5.12), $(\Omega_0 - \omega)/\Omega \ll n_\parallel^2$. Since for larger $n_\parallel$ particles can interact with the wave at frequencies further from resonance, this last condition means that our previous estimate of the range of angles over which relativistic effects matter will, if anything, be too narrow. An important effect which is immediately apparent from figure 5.3 is that towards the low field side (the left), there is a sharp boundary beyond which it is impossible to satisfy the resonance condition. This means that absorption is strictly zero beyond this point, whereas on the other side it decays gradually towards zero as the resonant velocities move outside the region in which there is a significant population of particles. In the non-relativistic case there is no such boundary and the absorption profile is much closer to being symmetrical about the resonance.

A calculation of the relativistic dielectric tensor elements was first given by Trubnikov (1959) and simplifications of the result appropriate to weakly relativistic plasmas were given by Dnestrovskii and Kostomarov (1962) and Shkarofsky (1966a,b). The calculation proceeds from the relativistic Vlasov equation which, linearised in the usual way, takes the form

$$\frac{\partial f_1}{\partial t} + \boldsymbol{v}\cdot\frac{\partial f_1}{\partial \boldsymbol{r}} - e(\boldsymbol{E} + \boldsymbol{v} \times \boldsymbol{B}_0)\cdot\frac{\partial f_1}{\partial \boldsymbol{p}} = e(\boldsymbol{E} + \boldsymbol{v} \times \boldsymbol{B})\cdot \frac{\partial f_0}{\partial \boldsymbol{p}} \quad (5.13)$$

where the unperturbed distribution $f_0$ is a function of $p_\parallel$ and $p_\perp$, the parallel and perpendicular momentum components, and the perturbation $f_1$ is a function of $\boldsymbol{r}$, $\boldsymbol{p}$ and $t$. For a thermal plasma, $f_0$ is proportional to $\exp(-\varepsilon/T)$, where $\varepsilon$ is the relativistic energy

$$\varepsilon = (m^2c^4 + c^2p^2)^{1/2}.$$

The calculation proceeds, in the same way as the familiar non-relativistic version, by integration along the unperturbed orbits which are given by

$$\frac{\mathrm{d}\boldsymbol{r}}{\mathrm{d}t} = \frac{\boldsymbol{p}}{m\gamma}$$

$$\frac{\mathrm{d}\boldsymbol{p}}{\mathrm{d}t} = -\frac{e\boldsymbol{p}}{m\gamma} \times \boldsymbol{B}_0.$$

Various alternative forms are possible for the end result, and a good review of the subject, which discusses their derivation in more detail than is possible here, has been given by Bornatici *et al* (1983). The dielectric tensor, calculated by integration along unperturbed orbits, is a fourfold integral over time and the three momentum components. If the integration over time is carried out, similar to the usual treatment of the non-relativistic case, an expression similar to (1.4) and (1.5) is obtained. Since this consists of a sum over different harmonics, it is most suited to calculations in which there is little overlap between the contributions from different harmonics, so that the contribution corresponding to the frequency of interest can be isolated. This is the case for comparatively low electron energies.

Alternatively, if the unperturbed distribution is isotropic, a transformation introduced by Trubnikov (1959) allows the integral over the angular variables to be carried out when the $p$ integration is expressed in spherical polars. The result is a double integral over time and $p$, which is more suitable for high temperature plasmas than the sum over harmonics.

Even at the temperatures anticipated in a reactor plasma, the plasma is weakly relativistic in the sense that

$$\mu = \frac{mc^2}{T} = \frac{c^2}{v_{th}^2}$$

is a large parameter. This makes it possible to expand in terms of the reciprocal of this parameter and simplify the dielectric tensor. This was done for perpendicular propagation by Dnestrovskii and Kostomarov (1961) and extended to arbitrary $n_\parallel$ by Shkarofsky (1966a,b). Shkarofsky, whose work has been widely used, shows that the dielectric tensor elements can be expressed in terms of the integrals

$$\mathcal{F}_q \left(n\mu\Omega_0/\omega, n_\parallel\right) = -\mathrm{i} \int_0^\infty \frac{\mathrm{d}\tau}{(Y(\tau))^q} \exp\left(\mu(1 - Y(\tau)) - \mathrm{i}n \frac{\mu\Omega_0}{\omega} \tau\right)$$

(5.14)

where

$$Y(\tau) = [(1 - \mathrm{i}\tau)^2 + n_\parallel^2\tau^2]^{1/2}$$

(5.15)

and the values $q = \frac{3}{2}, \frac{5}{2}, \frac{7}{2}, \ldots$ occur.

For perpendicular incidence these functions reduce to functions

$$F_q(z) = -\mathrm{i} \int_0^\infty \frac{\mathrm{d}\tau}{(1 - \mathrm{i}\tau)^q} \exp\left(\mathrm{i}z\tau\right)$$

(5.16)

which were introduced by Dnestrovskii *et al* and shown by Shkarofsky to be expressible in terms of the plasma dispersion function.

While these results reduce the problem of evaluating the dielectric tensor elements to that of carrying out a single integral, the integral is an awkward one, with a rapidly oscillating integrand. Much more recent work on the problem (Airoldi and Orefice 1982, Krivenski and Orefice 1983) has achieved further simplification by relating Shkarofsky's $\mathcal{F}_q$ functions to the plasma dispersion function, for which numerical evaluation routines are readily available. It has also been shown (Bornatici and Ruffina 1986, Orefice 1988) how more general distributions like anisotropic Maxwellians or loss-cone distributions can be treated in such a way that the dielectric tensor is again expressed in terms of Shkarofsky's functions. The mathematical properties of these functions, including various useful approximate forms, are discussed further by Robinson (1986, 1987).

We do not have space here for a detailed discussion of the relativistic dielectric tensor and the absorption properties of electron cyclotron waves which it predicts. We simply state some of the most important properties, attempting to explain the physics underlying them, and refer the reader who seeks more detail to the review article by Bornatici *et al* (1983), the book by Akhiezer *et al* (1975) or the papers of Fidone *et al* (1978, 1982) and Litvak *et al* (1977).

Bornatici *et al* quote a large number of analytic estimates of the absorption coefficients in various regimes. Different expressions are obtained depending both on the density and on the angle of propagation. As has already been discussed, the angle of propagation is important in determining whether relativistic effects are important, non-relativistic approximations being adequate sufficiently far from perpendicular. The dependence on density arises because the hot plasma part of the dielectric tensor always contains the factor $(\omega_p/\omega)^2$. At low densities, or at high harmonics, the correction to the hermitian part of the dielectric tensor is negligible, and the anti-hermitian part is a small correction, producing damping proportional to the density. At higher densities, the thermal contributions to the dielectric tensor can become comparable to the cold plasma tensor elements in the vicinity of the resonance, and the density dependence becomes more complicated.

At the fundamental the absorption of the O mode is greatest at perpendicular incidence and falls off rapidly away from perpendicular, while the X mode is only weakly absorbed near perpendicular but typically becomes strongly absorbed with an angle of incidence greater than 45° to the field. A qualitative understanding of this behaviour can be obtained if it is remembered that the absorption is due to either the right circularly polarised component of the electric field or the $\boldsymbol{v} \times \boldsymbol{B}$ force arising from the wave magnetic field and the electron velocity along the direction of the steady magnetic field. The former is the dominant absorption mechanism of the X mode for near-parallel incidence, and the right circularly polarised component falls off as the direction becomes more nearly perpendicular to the field. For the O mode the latter mechanism is dominant at perpendicular incidence, and as the angle changes the parallel field component decreases.

At the second harmonic the X mode is strongly absorbed at perpendicular incidence and over a wide range of angles, while the O mode is more weakly absorbed. The stronger absorption of the X mode at the second harmonic for propagation across the field occurs for the same reason as has already been discussed for ion cyclotron heating, where it was shown that the thermal corrections are larger than at the fundamental. Since the electron cyclotron wave phase velocity is of the order of the velocity of light, $k_\perp v_{th}/\Omega$ is around $v_{th}/c$. The thermal corrections are of the order of $(v_{th}/c)^4$ at the fundamental and of order $(v_{th}/c)^2$ at the second harmonic.

An interesting feature of the results is that in many cases the total absorption of a wave propagating across an inhomogeneous magnetic field in a tokamak-type geometry is predicted to be the same, whether the relativistic theory or a non-relativistic theory as given by, for example, Antonsen and Manheimer (1978, 1980) is used. The reason is most easily seen if we look at the weakly relativistic dielectric tensor elements in an alternative to the usual Shkarofsky form, given by Chu and Hui (1983) (see also Chu 1985). In the form given by these authors it is apparent that the weakly dielectric tensor elements are of basically the same form as the non-relativistic ones, with the usual resonant denominator

$$\omega - n\Omega - k_\parallel v_\parallel$$

replaced with

$$\omega - n\Omega/\gamma - k_\parallel v_\parallel$$

or, approximately,

$$\omega - n\Omega(1 - v^2/2c^2) - k_\parallel v_\parallel.$$

We now write the dispersion relation as

$$D_0(\omega, k) + iD_1(\omega, k) = 0$$

where $k$ is the component of the wavenumber across the field and the other wavenumber components, which are constant in a simple slab geometry, have been suppressed. Then the imaginary part of $k$ is given by

$$\operatorname{Im} k = k_i = -\frac{D_1(\omega, k_0)}{\partial D_0/\partial k_0}$$

where $k_0$ is the solution of $D_0(\omega, k) = 0$. Usually $D_0$ can be taken to be the cold plasma result, to a good approximation, while $D_1$ arises from a finite Larmor radius correction and is of the form

$$\operatorname{Im} \int \frac{H(\boldsymbol{v})}{\omega - n\Omega(1 - v^2/2c^2) - k_\parallel v_\parallel} \, d^3v$$

with $H(\boldsymbol{v})$ involving the electron distribution function and the usual Bessel functions. In the inhomogeneous system, the behaviour around a cyclotron resonance can usually be approximated by a linear gradient, so that we may take

$$\omega - n\Omega \approx \frac{n\Omega x}{L}$$

where L is a gradient scale length. The total absorption depends, within the WKB approximation, on

$$\int k_i \, dx$$

which is proportional to

$$\text{Im} \int dx \int \frac{H(\boldsymbol{v})}{n\Omega x/L - \frac{1}{2}n\Omega(v^2/c^2) - k_{\parallel}v_{\parallel}} \, d^3v.$$

Inverting the order of integration, and using the fact that the integral is taken around the singularity in the usual Landau way, we obtain

$$\text{Im} \int d^3v \int dx \frac{H(\boldsymbol{v})}{n\Omega x/L - \frac{1}{2}n\Omega(v^2/c^2) - k_{\parallel}v_{\parallel}} = \int d^3v \frac{L}{n\Omega} \pi H(\boldsymbol{v})$$

which is the same whether the relativistic correction is present or not.

A notable exception to this occurs in the case of the X mode at the fundamental. Here the integrals of the above form with $n = 1$ contribute to the cold plasma dielectric tensor in the limit of temperature going to zero. Thus, these resonant integrals are not, in this case, a small perturbation to the cold plasma result and the relativistic effects enter in a different way from that assumed above. As has been pointed out by Bornatici and Englemann (1980), this produces a quite different dependence on density of the total absorption in the relativistic and non-relativistic cases. Even if the relativistic and non-relativistic cases give the same result for the total absorption, it is important to emphasise once more that the absorption profile may well be quite different and that its correct calculation requires the relativistic form of the plasma dielectric tensor.

The calculation of absorption profiles and optical depths for electron cyclotron waves has generally been done on the basis of WKB theory, but, as in ion cyclotron heating, the wave propagation is complicated by the presence of Bernstein modes to which the X mode may couple, and cyclotron harmonic waves (Akhiezer *et al* 1975) to which the O mode couples. In the case of perpendicular incidence, the problem has been treated in terms of mode conversion by Antonsen and Manheimer (1978) and Cairns and Lashmore-Davies (1982, 1983). The transmission coefficient of the fast wave agrees with a WKB calculation, but mode conversion gives the possibility of reflection of some of the energy. For non-perpendicular incidence the Bernstein mode is damped. The order-reduction schemes described for ion cyclotron theory should also be applicable to this case, but less work has been done on this problem. The behaviour of the X mode and Bernstein mode dispersion curves has been analysed by Lazzaro and Ramponi (1981) and Bornatici *et al* (1981). Both of these studies show that above a critical density, which is such that $\omega_p$ is of the same order as $\Omega_e$, the behaviour of the dispersion curves is such that mode conversion would be expected to occur. In some recent work, the problem of absorption by more general distribution functions has been looked at. Maxwellians with drift velocities can be treated by Lorentz transformation of the zero drift velocity results, and some work has also been done on the distributions with different parallel and perpendicular temperature and on loss-cone distributions. Given such results, it will then, of course, be possible to deal with any distribution which is a superposition of these special distributions.

An interesting consequence of the relativistic effect on the resonance condition is that it may allow absorption by a wave whose frequency is below the electron cyclotron frequency everywhere in the plasma. The general idea, which was given by Fidone *et al* (1985, 1986), can be understood by referring to figure 5.3. We can see from this that the relativistic effect makes the resonant velocities, in the region where the cyclotron frequency is above the wave frequency, considerably less than they would be according to non-relativistic theory. If we now think of the point where the wave frequency equals the cyclotron frequency as not being in the plasma at all, but rather to be on the low field side of the plasma column, then it is still possible, if the temperature is large enough, for the wave to be absorbed effectively by particles on the tail of the distribution. Fidone *et al* show that an obliquely incident X mode can be absorbed efficiently at a downshifted frequency by this mechanism. The advantages of this technique are that it allows the use of waves at a lower frequency than is otherwise the case, or alternatively, if the technology is available to produce radiation at a given frequency, it allows a higher magnetic field to be used in the tokamak. Also, as we shall see in the next chapter, an essential feature of schemes for electron cyclotron current drive is that energy by given to particles in the tail of the distribution, so that downshifted frequencies are also of interest in this context.

This type of heating has been investigated experimentally on the PLT tokamak at Princeton. The results indicated that the wave was absorbed on electrons in a high energy electron tail produced by the DC electric field in the machine. The perpendicular velocity of these electrons was increased, and, because of the decrease in collision frequency with energy, this allowed the particles to be further accelerated by the DC field, the resulting high energy tail being detected by the increased electron cyclotron emission which it produced. This demonstrated that the idea worked in principle, though in a machine with a higher plasma temperature the intention would be that energy be absorbed by particles in the tail of the Maxwellian distribution and transfer energy to the bulk of the plasma through collisions, rather than go into creating a high energy tail.

## 5.4 Non-linear effects at high powers

Recently there has been some interest in heating at high power levels (Nevins *et al* 1987, Taylor *et al* 1988). For high power gyrotrons a regime might be reached in which some small corrections to the linear theory might be needed, but the main stimulus for development of theories for very high intensities has been the MTX experiment (Thomassen 1988) in which a free electron laser will be used to produce heating using pulses with peak power up to 8 GW and average power around 2 MW. Such peak power levels take us into a highly non-linear regime of electron dynamics, and both the above papers consider the single particle dynamics of electrons in very high amplitude electron cyclotron waves.

To obtain some insight into the electron behaviour, it is useful to consider an average of the equations of motion over a cyclotron orbit. In this way we obtain a description of the behaviour over a long timescale without the necessity of following all the details of the cyclotron motion. Both of the papers referred to at the beginning of this section adopt this approach, that by Nevins *et al* using an averaged Hamiltonian approach and that of Taylor *et al* using a Lagrangian method. We shall describe the second of these, using a Lagrangian formalism which may be unfamiliar to many readers, and which may be of more general use for constructing gyro-averaged equations. The basic idea of this method was obtained from the work of Littlejohn (1983, 1985).

We begin from what Littlejohn refers to as a phase space Lagrangian, defined in terms of the usual Hamiltonian by

$$L(q, p, \dot{q}, t) = p \cdot \dot{q} - H(p, q, t). \tag{5.17}$$

The Euler–Lagrange equations applied to the variational principle

$$\delta \int L \, dt = 0$$

with $p$ and $q$ allowed to vary independently give Hamilton's equations in their usual form. The advantage of this technique, which is also described in the well known text of Goldstein (1980), is that it gives the equations of motion in the first-order form of Hamilton's equations, but the variational principle can be used to give the equations of motion in terms of an arbitrary set of independent coordinates, not restricted to those obtained via a canonical transformation.

Applying this to the motion of an electron in an electromagnetic field, we have

$$p = \gamma m \boldsymbol{v} - e\boldsymbol{A}$$

$$= m\boldsymbol{u} - e\boldsymbol{A}$$

where $\gamma$ is the usual relativistic factor and $\boldsymbol{A}$ is the combined vector potential of the wave and the steady magnetic field. The Lagrangian of (5.17) is then

$$L = (m\boldsymbol{u} - e\boldsymbol{A}) \cdot \dot{\boldsymbol{r}} - (m^2 c^4 + m^2 c^2 u^2)^{1/2} + e\phi \qquad (5.18)$$

where $\phi$ is the scalar potential.

To illustrate the averaging procedure let us look at a simple case, that of a wave in the O mode propagating at right angles to a steady magnetic field of strength $B_0$. The vector potential can then be taken to be

$$\boldsymbol{A} = B_0 x \hat{\boldsymbol{y}} + A_1 \cos{(kx - \omega t)} \hat{\boldsymbol{z}} \qquad (5.19)$$

the gauge chosen so that the scalar potential of the wave vanishes. We substitute this into the Lagrangian, and on the assumption that the field is small enough that the motion is a small perturbation of the basic cyclotron motion, go to the guiding centre coordinates defined by

$$\boldsymbol{r} = \boldsymbol{R} + (u_\perp / \Omega) \sin \theta \hat{\boldsymbol{x}} - (u_\perp / \Omega) \cos \theta \hat{\boldsymbol{y}}$$

$$\boldsymbol{u} = u_\perp \cos \theta \hat{\boldsymbol{x}} + u_\perp \sin \theta \hat{\boldsymbol{y}} + u_\parallel \hat{\boldsymbol{z}}$$

$$\dot{\boldsymbol{r}} = \dot{\boldsymbol{R}} + (\dot{u}_\perp / \Omega) \sin \theta \hat{\boldsymbol{x}} + (u_\perp \dot{\theta} / \Omega) \cos \theta \hat{\boldsymbol{x}}$$

$$- (\dot{u}_\perp / \Omega) \cos \theta \hat{\boldsymbol{y}} + (u_\perp \dot{\theta} / \Omega) \sin \theta \hat{\boldsymbol{y}}.$$

The procedure is now to make this coordinate transformation in (5.18) and (5.19) and average the Lagrangian over the rapidly varying angle $\theta$. If the wave is taken to be at the fundamental, this is done with the assumption that

$$\Phi = \theta - \Omega t$$

is a slowly varying quantity. Also, when the guiding centre coordinates are substituted into (5.19) we make the usual expansion in series of Bessel functions.

The result of all this is an averaged Lagrangian, to which the variational principle is applied in order to obtain equations for the time dependence of the guiding centre coordinates, over timescales long compared with the cyclotron period. For the example described above, the two equations (out of the total of six) which will be of interest to us are

$$\dot{u}_\perp = \tfrac{1}{2}(\Omega_1 \dot{z} \sin \Phi) \tag{5.20}$$

and

$$\dot{\Phi} = \Omega/\gamma - \omega + \tfrac{1}{2}(\Omega_1 \dot{z}/u_\perp) \cos \Phi \tag{5.21}$$

in obtaining which we use the small amplitude expansion of the Bessel functions. The quantity $\Omega_1$ is equal to $ekA_1/m$. The same result has been obtained by Suvarov and Tokman (1983) using direct averaging of the equations of motion. The advantage of the Lagrangian method is that it provides a systematic technique which can be extended, in a comparatively straightforward way, to arbitrary directions of propagation and wave polarisation.

Assuming that the velocity along the field is constant, the behaviour of the solutions of (5.20) and (5.21) is most easily seen by means of a phase space diagram for this system, a typical example of which is shown in figure 5.4. The equations have an elliptic fixed point at $\Phi = \pi/2$ and at a value of $u_\perp$ a little above that for which $\omega = \Omega/\gamma$. The value of $u_\perp$ at this fixed point increases with wave amplitude. At $\Phi = -\pi/2$, there are two fixed points at small enough wave amplitudes, a hyperbolic one at values of $u_\perp$ a little below that for which $\omega = \Omega/\gamma$ and an elliptic one at a low velocity. If the wave amplitude is made big

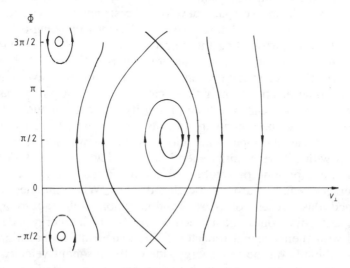

**Figure 5.4** Phase space trajectories of the average electron motion in a high amplitude electron cyclotron wave.

enough, these last two fixed points approach each other and finally disappear by merging. The whole diagram is, of course, periodic in $\Phi$. Similar phase space trajectories were obtained by Nevins *et al* (1987) from their averaged Hamiltonian analysis.

The physics underlying the periodic orbits surrounding the elliptic fixed point can be understood as follows. Take a particle which satisfies the cyclotron resonance condition exactly, and whose phase with respect to the wave is such that it gains energy. As it does so its cyclotron frequency decreases, because of the relativistic mass shift, and there will be a drift in the phase relation between the particle and the wave. Eventually the phase reaches a value where the particle begins to lose, rather than gain, energy. The frequency will then increase again and the particle will come back into resonance, losing energy all the while. As its energy goes below the value at which it is exactly in resonance, its cyclotron frequency goes above the wave frequency. Again there is a shift in the phase, until it reaches a point where the particle starts to be accelerated again and the whole cycle repeats itself.

If the electrons pass through a beam with a smooth intensity profile, Nevins *et al* describe the following heating process for large amplitude waves. Assume that the change in intensity profile is on a length scale long enough that we can make an adiabatic assumption, that is we can regard the electrons as moving along phase space orbits as shown in figure 5.4, but with the phase space changing slowly. If we now think of the behaviour of electrons close to the resonant velocity for which $\omega = \Omega/\gamma$, then before they enter the beam they are of course on phase space trajectories which are simply defined by $u = \text{constant}$. On going into the beam the phase space develops the structure described above, with the trajectories for almost resonant electrons pinching-off to form the closed loops surrounding the elliptic fixed point. It is assumed that initially most electrons are on the low energy side of the resonant velocity. As the electrons move into the beam and the wave amplitude increases, some electrons enter these closed regions of phase space and are trapped. If the beam width and intensity are large enough, then the electrons can describe a number of orbits around the elliptic fixed point, before becoming untrapped once more as they move out of the beam. Electrons with different initial energies will follow different trajectories, with different periods of rotation about the fixed points, resulting in their being spread out around the closed loops. When the electrons are untrapped this process of phase randomisation will lead to approximately half going onto the trajectory on the high energy side which merged with their original trajectory. The end result is that a group of electrons initially on the low energy side of the resonant velocity is split into two energy groups, one with the original energy, and one with an energy above that of resonant particles. This produces an effective heating mechanism. Nevins *et al* have described elaborations of this mechanism, in which the parameters of the system are varied so that as the particle goes through the beam, the regions of phase space in which particles are trapped move towards higher energies. This mechanism, which they describe as trapping in phase space 'buckets', may provide an effective way of moving substantial numbers of electrons to higher energies.

## 5.5 Experimental aspects

As mentioned in the introduction, the power available for electron cyclotron resonance heating has been limited until comparatively recently, and to some extent still is, by the high frequencies involved. At the fundamental the relation between the frequency and the tokamak magnetic field is 28 GHz per tesla. However, there has been a continuous development in gyrotrons towards both higher frequencies and higher powers, so that a continuous wave power of 1 MW at around 100 GHz now seems to be a possibility in the near future. Free electron lasers may also provide a source of very high intensity pulses in the frequency range of hundreds of gigahertz.

Since the mechanism of gyrotrons is closely related to that of electron cyclotron resonance heating, it may be of interest to digress from our main topic to give a brief account of their basic principles (for more details see e.g. Kreischer *et al* 1985, or Read and Granatstein 1985). Essentially the gyrotron consists of a beam of electrons spiralling around the lines of an externally produced magnetic field, amplifying an electromagnetic wave which has an electric field component transverse to the steady magnetic field. The mechanism of amplification depends on a phase bunching of electrons. In the frame of reference in which the axial velocity of the electrons is zero, the wave frequency is a little above the cyclotron frequency. Initially, electrons have random phase in their cyclotron orbits and are equally likely to gain or to lose energy from the wave field. However, an electron which gains energy experiences a lowering of its cyclotron frequency because of the relativistic mass shift and so its phase slips relative to that of the wave. On the other hand, if an electron loses energy then its cyclotron frequency moves closer to the wave frequency, so it remains in the same phase relationship to the wave and continues to lose energy. The net result is a bunching of the electron phases in their cyclotron orbits in a position where they lose energy to the wave, producing an instability which feeds energy into the wave. Saturation of this cyclotron maser instability occurs when the particles lose so much energy that their cyclotron frequencies go above the wave frequency and they go out of phase again. Clearly this is very similar to the behaviour described in the previous section, and involves the same sort of phase space dynamics. Conversion of electron energy to wave energy can take place with an efficiency of around 50%.

Detailed experiments on the use of electron cyclotron waves for plasma heating began in the mid 1970s with the TM-3 machine (Alikaev *et al* 1976), using an almost circularly polarised beam incident from the low field side. The results of this and other similar experiments (see Bornatici *et al* (1983) for a summary of experiments up to that time and detailed references) were essentially as expected from theory, though the small size of early devices made accurate, quantitative comparison difficult, since absorption was not complete on a single pass and multiple reflections from the walls occurred. Absorption in TM-3 was found to occur at the fundamental and second harmonic, with the resulting heating localised around the cyclotron resonance. The absorption led to heating of the bulk of the plasma with no significant generation of highly energetic electrons. The electron confinement time was found to increase with electron temperature, and no significant increase in radiated power was noted during the pulse. This latter observation was taken to indicate that there was no sizeable population of hot electrons produced by the heating, nor any substantial influx of impurities.

The ISX-B tokamak (Eldridge *et al* 1980) used microwaves at 35 GHz, injected from the midplane of the high field side at an angle of 45° to the magnetic field. The radiation was unpolarised and calculations indicated that the component in the X mode should be almost totally absorbed, but that absorption of the O mode should be incomplete. The electron temperature was found to rise linearly with power, reaching about 1.4 keV, as compared to the Ohmic temperature of around 1 keV, at a power of 90 kW. It was also found that the temperature profile became more peaked and the electron density increased by around 15%.

With the T-10 tokamak (see for example Alikaev and Parail 1984) there was a substantial increase in the available power to around 1 MW, provided by a set of four gyrotrons whose output was launched from the low field side and focused at the plasma centre. The frequency was around 80 GHz, corresponding to a fundamental resonance at a field of 3 T. Experiments were carried out both with this toroidal field and with 1.5 T, the latter giving a second harmonic resonance. The experiment at the second harmonic resonance was again in a regime where the wave was not expected to be absorbed in a single pass, but when the field was increased in order to produce absorption at the fundamental, single pass absorption was obtained for the first time. In this latter regime the heating efficiency was as much as 90%. An interesting result of this experiment was that the heating efficiency was not strongly affected by a shift of the resonance position away from the centre of the plasma column by quite a significant fraction of the minor radius. This means that the plasma can be heated, even though the density at the centre of the machine is somewhat above the density limit for accessibility of the wave.

More recent experiments (for a summary see for example Riviere 1987) have indicated that the effect on energy confinement of electron cyclotron heating is dependent on the plasma density. At lower densities ($n_e < 10^{19}$ m$^{-3}$) there is a significant improvement in energy confinement. These densities are such that collisions are not strong enough to maintain a Maxwellian distribution and it is thought that the resulting population of hot electrons may produce an improvement in plasma stability. At intermediate densities the distributions remain close to Maxwellian and the energy confinement time generally reduces with temperature, with the same kind of scaling as has been observed for other heating schemes (Kaye and Goldston 1985). However, some observations have shown that a narrow peak in electron temperature may be produced if the resonance is placed exactly on the magnetic axis of the tokamak. These suggest that energy confinement close to the axis is not very much affected by the heating (e.g. Cavallo *et al* 1985), so that energy deposited on-axis produces a steep temperature gradient. At high densities ($n_e > 5.5 \times 10^{19}$ m$^{-3}$) experiments in the Doublet III tokamak have shown that an improvement in the energy confinement time over that in the corresponding Ohmically heated plasma may be produced (Prater *et al* 1986) with the absorption region near the plasma edge.

Interesting results were obtained in Doublet III (Ejima and Prater 1987) using plasmas in which the density was above the limit at which the centre was accessible to the waves. Evidence from both soft x-ray emissions and the effect of the heating on the frequency of sawtooth oscillations suggested that the waves were indeed being absorbed near the edge of the plasma column. However, the temperature profile remained very similar to that obtained at lower densities, when the wave could reach the resonance on the axis of the plasma, and be absorbed. The lack of any change in the temperature profile was attributed by Ejima and Prater to a property of plasma transport in a tokamak, rather than to any anomaly in the propagation and absorption of the wave. Similar effects had already been seen in neutral beam heated tokamaks, and had given rise to the concept of 'profile consistency' (Furth *et al* 1985), the essential idea of which is that the electron temperature profile has a shape imposed by constraints of stability to resistive kink modes, and takes up this shape regardless of the energy deposition profile. Transport processes within the plasma, the details of which are still poorly understood, are presumed to act in such a way as to give rise to this preferred density profile. Another question which arose in these experiments was the mechanism of absorption of the electron cyclotron waves, since the cyclotron resonance was at the centre of the plasma column so that waves would have to be absorbed away from the resonance. Calculations suggested that before being fully absorbed the waves would have to make a number of reflections off the wall of the tokamak, but the amount of power being lost to the walls was small. This indicated that the plasma was absorbing the wave more effectively than was expected from the calculations of linear wave damping. An explanation put forward to explain the higher level of damping was the occurrence of a parametric instability of a type suggested by Matsuda (1986). Since the centre of the plasma column is not accessible to the wave, it is refracted in such a way as to travel almost parallel to the external field for some distance. Matsuda shows that in these circumstances parametric decay to an ion sound wave can occur, and that the threshold and growth rate of this instability are such that it might be expected to be excited in the Doublet III experiment.

At the beginning of the RF heating pulse, a feature of electron cyclotron heating appears to be a marked decrease in density of up to 50%. During a time of the order of the electron collision time, the density profile broadens and particles are lost from the system, presumably because the heating is having some effect, not yet understood, on the plasma transport processes.

As well as their use as a plasma heating system to produce the sorts of temperatures required for ignition in reactor plasmas, various other uses have been suggested for electron cyclotron waves. For instance, electron cyclotron waves may be used to create and heat the plasma in the initial stages of a discharge, with a view to reducing the loop voltage and volt-seconds required of the Ohmic heating coils (Peng *et al* 1978). Although the loop voltage around a tokamak is only of the order of 1 V when the plasma is at the operating temperature, it may need to be of the order of several hundred volts during the start-up phase of a reactor-sized machine. Any way of reducing this will make a substantial contribution to economising on the requirements of the power supply system. This is particularly important in view of the so-called radiation barrier which occurs at temperatures below about 100 eV (Hawryluk and Schmidt 1976). In this regime the plasma energy balance is very much affected by radiation from low-$Z$ impurities, the high losses from this effect putting severe conditions on a reactor regime Ohmic heating system. The scenario envisaged by Peng *et al* (1978) is that in the initial stages, when the electron density is low, there will be strong absorption of the fundamental X mode at the cyclotron resonance. This happens because at sufficiently low electron densities in a partially ionised plasma the screening out of the right-hand circularly polarised component is not complete and there is relatively strong absorption, leading to further ionisation and an increase in the plasma density. As the density increases the screening increases and absorption at the cyclotron resonance falls off. The wave, incident from the low field side, is then expected to pass through the cyclotron resonance and reach the upper hybrid layer where it will be converted to a Bernstein mode. The calculations of Peng *et al* suggest that quite modest microwave power levels used to irradiate a cold plasma in the X mode from the high field side may reduce the loop voltage required during start up by a factor of five to ten.

The localised and controllable nature of the energy absorption region of electron cyclotron waves means that it may be possible to use them to control the temperature and current profiles with a view to enhancing the stability of the plasma. Chan and Guest (1982) analyse a scheme in which heating of the plasma around the $q = 1$ and $q = 2$ surfaces leads to a flattening of the temperature profile and of the current profile (the temperature dependence of resistivity means that the current density is highest where the temperature is highest). This produces, in turn, a stabilisation of tearing modes, which are driven by gradients around rational $q$ surfaces. Such modes are thought to be associated with disruptions, which cause loss of plasma equilibrium and rapid destruction of the discharge. The analysis of Chan and Guest shows that it is quite feasible, using reasonable microwave power levels, to produce the required flattening of the profile and that this is likely to be an attractive way of achieving disruption-free operation of a tokamak.

The final main use of electron cyclotron waves is in current drive, where obliquely incident waves may be used to produce a current, with the object of reducing, or even eliminating entirely, the need for inductively driven current in a tokamak. This will be discussed in the following chapter.

# 6  Current Drive

## 6.1 Introduction

In previous chapters we have discussed wave propagation and absorption in various frequency ranges and have shown how radiofrequency waves may be used to heat a plasma. In addition to using radiofrequency waves simply to heat the plasma, there has also been a great deal of interest in their use to drive current around a tokamak. Conventional ohmic current drive relies on inducing an EMF around the toroidal plasma by changing the magnetic flux which passes through its centre. However, it is clear that the magnetic flux cannot change monotonically for an indefinite period of time and that the process is necessarily pulsed in nature, though in a reactor the loop voltage required is small and the pulses could be quite long, perhaps of the order of an hour. From the engineering point of view it would be much simpler to have a reactor which could operate in a continuous mode, and it is this which has motivated research into various schemes for producing non-inductive current drive. Even if this is not practicable, radiofrequency current drive may still be useful in modifying the current profile in order to suppress magnetohydrodynamic instabilities. There is little scope for changing the current profile produced by Ohmic heating, since it is determined by the transport properties of the plasma. On the other hand, the region in which waves are absorbed is controllable to a considerable extent, so that localised generation of current is possible.

In this chapter we discuss some of the main ideas involved in radiofrequency current drive. The essential feature is that it is necessary for the waves to be absorbed in such a way that some kind of asymmetry is produced with respect to the toroidal direction. An examination of exactly how this works will involve us in a more detailed consideration of the balance between wave absorption processes and collisions than was necessary for examining heating, where the only real consideration is to get the energy into the centre of the plasma and absorbed there. Efficient current drive will require that we be able to control just how the absorption takes place, since good efficiency depends on exciting the right part of the particle distribution function. Before going on to a detailed discussion of RF current drive we might mention that the other main contender for current drive is neutral beam injection. An excellent review of current drive theory has recently been given by Fisch (1987) and we refer the interested reader to this for further details.

## 6.2 Basic principles

As shown in section 1.4, it is possible for waves to interact with particles through a Landau resonance, in which the particle is accelerated in the direction parallel to the magnetic field, or a cyclotron resonance, in which energy is mainly put into the perpendicular degree of freedom. At first sight it might appear that only the first of these would be of any use for current drive, but it was shown by Fisch and Boozer (1980) that this is not so and that, at least in principle, cyclotron resonance is almost as efficient as Landau resonance in driving current.

The figure of merit generally used to characterise the current drive efficiency is the current density per unit absorbed power density. The following simple considerations lead to an appreciation of how it scales with the parameters of the plasma. If we consider first electrons pushed in the parallel direction, then a change in parallel velocity $\delta v_\parallel$ produces a current

$$j = q \delta v_\parallel$$

while the change in energy is

$$E = m v_\parallel \delta v_\parallel.$$

If the collision frequency for the electrons is $v$, then the current may be regarded as persisting for a time given by its reciprocal. To maintain a steady current the electron must be given a push at time intervals $1/v$, and so the power requirement per electron is

$$P = m v_\parallel v \delta v_\parallel$$

and we arrive at an estimate of the efficiency

$$\frac{J}{P_d} = \frac{q}{m v_\parallel v}. \tag{6.1}$$

Now, $v$ depends on the electron energy, going as $1/v^3$. From (6.1) we can identify two favourable regimes for current drive. Firstly, if $v_\parallel$ is small (less than the thermal velocity), then the velocity dependence of $v$

will mostly come from the perpendicular velocity, typically of order $v_{\mathrm{th}}$ and we see from (6.1) that

$$\frac{J}{P_{\mathrm{d}}} \sim \frac{1}{v_{\parallel}}. \tag{6.2}$$

On the other hand, if $v_{\parallel}$ is large we have $v \approx v_{\parallel}$ and so

$$\frac{J}{P_{\mathrm{d}}} \sim v_{\parallel}^2. \tag{6.3}$$

From this it is apparent that that there is a favourable scaling in both the small and large $v_{\parallel}$ limits. The first of these was the basis of early current drive schemes based on absorption of Alfven waves with low parallel phase velocity (Wort 1971), while the second was identified by Fisch (1978). As we shall see, the second of these regimes, reached through the use of lower hybrid waves, has been very successful, while the first has received little attention until recently. This is because in a tokamak it is necessary to take account of particle trapping in magnetic wells, produced by the weaker magnetic field towards the outside of the torus. Particles with small parallel velocities are precisely those which are trapped and this may be expected to provide a serious obstacle to current drive utilising such particles (Bickerton 1972).

A recent revival of interest in current drive using low frequency waves is based on the possibility of non-resonant damping described by fluid equations, rather than on resonant damping on slow particles (Ohkawa *et al* 1988, Mett and Tataronis 1989, Taylor 1989). The basic idea is that, while in an undamped Alfven wave the average value of $J \times B$ is zero, dissipation alters the phase difference between $J$ and $B$ so as to give rise to an average force which can drive a current. It is pointed out by Mett and Tataronis and by Taylor that viscous and resistive dissipation produce opposing phase shifts, which tend to cancel each other out.

The process can be analysed in terms of the transfer of helicity from the wave to the plasma. The helicity density is defined to be $A \cdot B$, where $A$ is the vector potential in some suitable gauge and $B$ is the magnetic field. A current with a component parallel to a magnetic field, as in a tokamak, involves non-zero helicity. A circularly polarised Alfven wave also carries helicity. Taylor (1989) points out that the interpretation of the process as a straightforward transfer of helicity from the wave to the current can be misleading, since the wave helicity can also be dissipated through the fluctuating current components. At the time of writing this whole field is a rather new development, and it is yet to be seen whether it can yield the high current efficiencies which its proponents claim.

If the wave–particle interaction is such as to increase the perpendicular degree of freedom of the particle, generation of a current depends on the velocity dependence of the collision frequency. Let us imagine two electrons with the same perpendicular energy and equal and opposite parallel velocities; then consider the perpendicular velocity of the one going to the right, say, to be increased. The collision frequency between the particle moving to the right and the ions is less than that of the other particle, so it loses momentum to the ions at a slower rate. The result is a net transfer of momentum between electrons and ions, so that the electrons have a drift velocity to the right and the ions to the left. Note that the wave imparts no momentum along the field lines to the particles in this scheme, it merely produces an asymmetry in the electron–ion drag which gives rise to a redistribution of momentum within the plasma.

To compare the efficiency of this process with the result obtained above for excitation of the parallel degree of freedom let us make a similar crude estimate. The rate of transfer of momentum is

$$\frac{dp}{dt} \approx m v_\parallel v$$

and in the steady state we may expect the change in $v$ to be accompanied by a change in $v_\parallel$ sufficient to bring the rates of momentum transfer in the two directions into balance once again. Thus

$$m v_\parallel v \approx m(v_\parallel + \delta v_\parallel)(v + \delta v)$$

or

$$\delta v_\parallel \approx -\frac{v_\parallel \delta v}{v}.$$

Again, the power required to maintain the change in $v$ continuously is

$$P_d \approx m v_\perp v \delta v_\perp$$

and so

$$\frac{J}{P_d} \approx -\frac{q v_\parallel \delta v}{m v_\perp v^2 \delta v_\perp}.$$

Now, if $v \sim 1/v^3$, we have

$$\frac{\delta v}{v} = -\frac{3 v_\perp \delta v_\perp}{v_\parallel^2 + v_\perp^2}$$

and so

$$\frac{J}{P_d} \approx \frac{3 q v_\parallel}{m(v_\perp^2 + v_\parallel^2)v} \tag{6.4}$$

which is typically of the same order of magnitude as the result given in (6.1).

The ratio of current density to absorbed power density can be converted into the ratio of total current $I$ to total absorbed power $P$, which is the quantity of most immediate relevance to tokamak studies. To do this, consider a ring of major radius $R$ and cross section $A$. Then the current in this is $AJ$ and the power absorbed $2\pi RAP_d$. Thus we have

$$\frac{I}{P} \approx \frac{1}{2\pi R} \frac{J}{P_d}. \tag{6.5}$$

Since the collision frequency scales as $nT^{-3/2}$ (neglecting the weak parameter dependence of the Coulomb logarithm), we can see from (6.1) and (6.4) that $I/P$ scales as $T/Rn$ and so efficient current drive is favoured by hot machines and by low density. Although the efficiency appears to decrease with the size of the machine, there are other more beneficial features of size. In a large machine it is easier to arrange for all the energy to be absorbed in a single pass by particles on the tail of the distribution.

Numerical estimates of the ratio $I/P$ indicate that a substantial fraction of the output power from a reactor would have to be recirculated to drive the current. The exact value depends on the reactor design, but the power requirements are not so large that steady state current drive can be dismissed immediately as being a practical impossibility. On the other hand, thay are large enough to ensure that careful design and optimisation of any scheme is important. Approximate estimates of efficiency, as we have so far, while sufficient to give the important scalings, are not good enough to provide a basis for detailed comparison with experiment and reactor design, and a lot more detailed work has been done, some of which we go on to describe now.

## 6.3 Langevin equations

A more exact estimate of steady state current drive efficiency usually relies on looking for a solution for the electron distribution function in which the effect of collisions is balanced by that of the wave, so that the equation to be solved is

$$\frac{\partial f}{\partial t} = \frac{\partial f}{\partial t_{coll}} + \frac{\partial f}{\partial t_{wave}}. \tag{6.6}$$

The first term on the right-hand side is a collision operator, usually of the standard Fokker–Planck form, and the second is most commonly a quasi-linear diffusion term.

However, before we consider this approach it is useful and illuminating to consider an alternative approach using Langevin equations (Fisch and Boozer 1980, Fisch 1987). The Fokker–Planck equation describes the relaxation of the distribution function towards its thermal equilibrium Maxwellian form under the action of small-angle collisions. It contains terms involving first-order derivatives of the distribution function, which can be interpreted as friction forces and terms with second-order derivatives which are diffusion terms. The Maxwellian is the result of a balance between the friction and diffusion terms.

An equivalent way of describing a system of this type is by a set of Langevin equations (see e.g. Chandrasekhar 1943), which are simply the equations of motion of individual particles under random forces, whose statistical properties reproduce the same diffusion and friction as in the Fokker–Planck equation. If the Fokker–Planck equation is of the form

$$\partial f/\partial t = A \cdot \partial f/\partial v + \tfrac{1}{2} D : \partial^2 f/\partial v \partial v$$

then $A$ is the friction force and $D$ the diffusion tensor (a second-order tensor). The corresponding Langevin equations are

$$dv/dt = F(t)$$

where $F$ is a random force with the statistical properties

$$\langle F \rangle = -A$$

and

$$\langle F(t)F(t') \rangle = 2D\delta(t - t').$$

Generally, to solve the problem with this approach would require integration of a large number of particle orbits with different forces, randomly distributed in the correct way, and would not represent any significant saving over direct solution of the Fokker–Planck equation. However, it was pointed out by Fisch and Boozer (1980) that for high velocity electrons the diffusion terms are small compared to the friction terms, so that it is not necessary to consider a statistical ensemble of different forces to reproduce the correct behaviour. Only the average slowing down need be taken into account. The next few paragraphs reproduce the arguments put forward in their paper.

Assuming azimuthal symmetry around the magnetic field direction, the electron distribution function depends on two variables, which it is convenient to take as $v$, the total velocity, and $u$, the velocity component along the magnetic field, both normalised to the thermal velocity. Then the frictional terms in the high velocity limit of the Fokker–Planck equation are reproduced by the Langevin equations

$$\frac{dv}{dt} = -\frac{v}{v^3} v$$

$$\frac{du}{dt} = -(2 + Z)\frac{v}{v^3} u. \tag{6.7}$$

Here $v$ is the usual collision frequency for a thermal particle given by

$$v = \frac{ne^4 \ln \Lambda}{4\pi\varepsilon_0^2 m^2 v_{th}^3}.$$

In (6.7) the $1/v^3$ dependence of the collision frequency appears in an obvious way. The second equation involves $Z$, the ion charge number. The difference in slowing down rate between $u$ and $v$ is a consequence of the fact that $u$ decreases not only because of the decrease in total velocity, but also because of pitch angle scattering, which changes the direction of the velocity. The result is that the velocity component in a particular direction decreases more rapidly than the total velocity. The $Z$ dependence of the rate of decrease of $u$, but not of $v$, is accounted for by the fact that in effect the ions are at rest and infinitely massive, so an electron scattering from an ion has the direction, but not the magnitude, of its velocity changed.

We now use the set of equations (6.7) to calculate the increment in current produced by a wave which changes the initial values of $u$ and $v$. This increment is calculated over the whole history of the particle which, according to (6.7), eventually ends up with zero velocity. Clearly this does not occur in practice, and the limit of zero velocity is to be thought of as incorporation of the particle into the bulk of the Maxwellian distribution, at which point collisional diffusion becomes important. The approximation may be expected to be a good one if the initial velocity is well above thermal, and the utility of the method depends on the fact that current is carried most effectively by fast particles, to which the equations are applicable.

To solve the equations note that

$$\frac{du}{dv} = (2 + Z)$$

so that if $u_0$, $v_0$ denote the initial values

$$\frac{u}{u_0} = \left(\frac{v}{v_0}\right)^{2+Z}. \tag{6.8}$$

The time-integrated current carried by the particle is

$$\int_0^\infty j\,dt = qv_{th}\int_0^\infty u\,dt = -qv_{th}\int_0^{u_0} \frac{u}{du/dt}\,du.$$

Using the second of (6.7) and eliminating $v$ through (6.8) this becomes

$$\frac{qv_{th}}{v(2+Z)}\int_0^{u_0} v_0^3\left(\frac{u}{u_0}\right)^{3/(2+Z)}du = \frac{qv_{th}}{v(5+Z)}v_0^3 u_0.$$

If we now think of a continuous process pushing electrons at $u$, $v$ to $u + \delta u$, $v + \delta v$, the ratio of current density to absorbed power density is

$$\frac{J}{P_{\rm d}} = \frac{q}{v(5+Z)v_{\rm th}} \frac{\delta(v^3 u)}{\delta(\frac{1}{2}mv^2)} \tag{6.9}$$

where for convenience we have dropped the subscripts 0. This holds for electrons interacting with the wave only at some isolated point in velocity space. For a wave–particle interaction described in terms of the rate of change of distribution function produced we have

$$\frac{J}{P_{\rm d}} = \frac{2q}{mv(5+Z)v_{\rm th}} \frac{\int v^3 u \left(\frac{\partial f}{\partial t}\right)_{\rm wave} d^3 v}{\int v^2 \left(\frac{\partial f}{\partial t}\right)_{\rm wave} d^3 v}. \tag{6.10}$$

From (6.9) we see again the conclusion we arrived at before, namely that it is almost as efficient to accelerate particles in the perpendicular direction as in the parallel direction. In the first case we change only $v$, while in the second we change both $u$ and $v$. The formula (6.9) was shown by Karney and Fisch (1981) to give good agreement with numerical solutions of the Fokker–Planck equation.

## 6.4 Adjoint methods

An alternative approach to the calculation of the plasma response to a given wave excitation term is through adjoint techniques (Antonsen and Chu 1982, Taguchi 1983). This method has proved to be more capable of generalisation than the Langevin equation method, in part because such techniques had already been developed in the problem of neoclassical transport (see e.g. Hinton and Hazeltine 1976). In order to illustrate the basic idea of this method let us consider the steady state problem

$$\frac{\partial f}{\partial t}\bigg|_{\rm coll} + \frac{\partial f}{\partial t}\bigg|_{\rm wave} = 0 \tag{6.11}$$

with

$$\frac{\partial f}{\partial t}\bigg|_{\rm coll} = v\left(\frac{1}{v^2}\frac{\partial f}{\partial v} + \frac{(1+Z)}{2v^3}\frac{\partial}{\partial \mu}(1-\mu^3)\frac{\partial f}{\partial \mu}\right) \tag{6.12}$$

$$= vCf$$

where $\mu = v_\parallel/v$ and again we take the velocities normalised to the thermal velocity. This collision term is the high velocity limit of the Fokker–Planck equation. It lacks many of the important properties associated with the exact collision operator, like energy and momentum conservation. However, it is a reasonable description of particles in a high velocity tail, and provides an analytically tractable example which will enable us to demonstrate explicitly the equivalence of the Langevin and the adjoint methods.

To obtain the Langevin equations corresponding to (6.12) we first note that there is no second-order derivative with respect to $v$, so that there is no diffusion in $v$. This means that $\langle v \rangle = v$ and

$$dv/dt = -v/v^2.$$

There is diffusion in $\mu$, but since $v$ is not a stochastic variable, it is possible to take the average in the friction term for $\mu$ to obtain

$$d\langle \mu \rangle/dt = -(1 + Z)\langle \mu \rangle/v^3.$$

Since we now have a closed pair of equations and the current depends just on the average value of $\mu$, we do not have to worry about the diffusion of this quantity. Introducing an average parallel velocity defined by

$$u = \langle \mu \rangle v$$

we regain the Langevin equations discussed in the last section. Having demonstrated that the problem posed here is exactly equivalent to the one we have already solved using the Langevin equations, we now show how the adjoint method can be used to obtain the same result.

If we define a scalar product of two functions in velocity space by

$$[f, g] = \int_0^\infty v^3 \, dv \int_{-1}^1 d\mu \, fg \tag{6.13}$$

then with $C$ the operator defined in (6.12) we have

$$[f, Cg] = [g, C'f] \tag{6.14}$$

where $C'$ is the adjoint operator given by

$$C'g = -\frac{1}{v^3} \frac{\partial}{\partial v} (vg) + \frac{(1 + Z)}{2v^3} \frac{\partial}{\partial \mu} \left( (1 - \mu^2) \frac{\partial g}{\partial \mu} \right). \tag{6.15}$$

We now use the relations (6.13)–(6.15) to calculate response functions giving the current driven and the power absorbed for a given $(\partial f/\partial t)_{\text{wave}}$.

If we take $g = v^3 \mu$ then we have

$$C'g = -(5 + Z)\mu$$

and from (6.14) and (6.12)

$$[f, C'g] = -(5 + Z) \int_0^\infty dv \int_{-1}^1 d\mu \, \mu v^3 f$$

$$= -\frac{(5 + Z)}{2\pi q v_{\text{th}}} J$$

$$= [g, \, Cf]$$

$$= -\frac{1}{v} \int_0^\infty dv \int_{-1}^1 v^6 \mu \left( \frac{\partial f}{\partial t} \right)_{\text{wave}} d\mu.$$

Thus

$$J = \frac{2\pi q v_{\text{th}}}{(5 + Z)v} \int_0^\infty dv \int_{-1}^1 d\mu \, v^6 \mu \left( \frac{\partial f}{\partial t} \right)_{\text{wave}}. \qquad (6.16)$$

The power density dissipated by collisions is

$$P_{\text{d}} = -v_{\text{th}}^2 \int \frac{1}{2} m v^2 \left( \frac{\partial f}{\partial t} \right)_{\text{coll}} d^3 v$$

$$= v_{\text{th}}^2 \int \frac{1}{2} m v^2 \left( \frac{\partial f}{\partial t} \right)_{\text{wave}} d^3 v. \qquad (6.17)$$

Combining (6.16) and (6.17) reproduces the result of the last section.

This type of technique has the advantage that it can be generalised to toroidal geometry (Antonsen and Chu 1982, Taguchi 1983) and to other problems such as the production of runaway particles in RF current drive (Fisch and Karney 1985). This is because the solution of the Fokker–Planck equation and its adjoint have already been studied in detail in neoclassical transport theory, which deals with the effects of toroidal geometry on transport theory and involves an expansion in terms of the inverse aspect ratio, that is, the ratio of the minor to the major radius. The essential trick in using this method is, as in our example above, to choose the inner product suitably then to choose the equivalent of our function g so as to give the required result. Antonsen and Chu showed that in toroidal geometry, and with a more exact form of the Fokker–Planck equation than we have used, the required function is that which describes the response of the plasma to a steady electric field. This, of course, is a problem which had already received a great deal of attention and had yielded results which could be applied directly to the current drive problem. Our example illustrates the essential mathematical structure of the method without introducing any of the complexities of neoclassical theory.

An interesting problem associated with radiofrequency current drive is that of runaways, which has been discussed by Karney and Fisch (1986). Runaway electrons occur when a large enough DC electric field is applied to a plasma (Dreicer 1960), simply because of the way in which the collisional friction term falls off with velocity. If an electron has a large enough initial velocity, the electric field force on it will outweigh the frictional force and it will be accelerated, causing the friction to become even smaller. This process can lead to the establishment of a population of high energy electrons. Its importance in connection with current drive lies in the fact that the radiofrequency field increases the energy of some of the electrons, so that if a DC electric field is also present, the population of runaway particles may be increased. Karney and Fisch pay particular attention to the problem of current ramp-up using lower hybrid waves. The essential idea of this is to increase the current using the radiofrequency waves. Even if no external DC field is imposed on the plasma, the toroidal system has a large self-inductance, so that any attempt to increase the current immediately leads to an induced field which opposes the change. As particles are accelerated along the magnetic field direction, they are pitch-angle-scattered by collisions, leading to an enhanced population of high energy electrons right around the bulk distribution, and not just in the direction of the initial acceleration. Those electrons which are scattered around the distribution and caused to move in the opposite direction to that in which they were initially accelerated may be picked up by the opposing DC field and become runaway electrons in the opposite direction to the electron current produced by the waves. This process obviously leads to a reduction in the efficiency of current drive when the current is being ramped-up. For details of the calculation, which involves use of the adjoint technique with a DC electric field included in the Fokker–Planck equation, we refer the reader to the original paper of Karney and Fisch. The topic of current ramp-up, and the way in which the radiofrequency-driven current interacts with the circuit properties of the plasma and the toroidal field coils is one to which we shall return later.

We should also mention at this point that the problems which we have discussed as being solved through the adjoint method or by means of Langevin equations can also be solved by more direct numerical methods. Examples of this treatment can be found in the work of Cordey *et al* (1982) and Ferreira *et al* (1984). These authors use an expansion of the pitch angle dependence of the distribution function in Legendre polynomials, reducing the flux surface averaged Fokker–Planck equation in tokamak geometry to a set of integro-differential equations in one variable, which are solved numerically. All these techniques give the same result, of course, to within the errors of numerical calculations and possible small differences arising from the use of different approximations to the Fokker–Planck equation.

The main limitation of the adjoint method, or the equivalent Langevin method, is the fact that the wave driving term $(\partial f/\partial t)_{\text{wave}}$ must be known. However, consideration of the standard quasi-linear term shows immediately that in general this depends on the distribution function $f$, which is not given by these techniques. Generally, in applying these methods, it is necessary to assume some distribution function. On the positive side, the Langevin theory shows that the ratio of current to absorbed power depends only on what part of velocity space is driven by the waves, assuming that the bulk of the distribution, which determines the collision frequency, is not changed. If this is known reasonably exactly then the ratio does not depend on the details of the tail of the distribution function. Although the analytic results we have given here have been obtained using an approximation to the Fokker–Planck equation which is only valid in the high energy limit, numerical calculations have shown that they are a reasonable approximation over a wide range of parameters (Karney and Fisch 1981).

A review of numerical methods used to solve the Fokker–Planck equation has been given by Karney (1986). Since both the collision term and the wave diffusion term conserve particles, the equation takes the form

$$\partial f/\partial t + \nabla \cdot S = 0$$

where the divergence operator is in velocity space and $S$ is a vector representing the flux of particles in velocity space under the combined effect of collisions and waves. A plot showing this flux can be a useful guide to understanding the behaviour of the plasma.

In a strong magnetic field the distribution function has cylindrical symmetry around the velocity axis parallel to the magnetic field, so it is most conveniently expressed in terms of cylindrical or spherical coordinates. A technique which has been widely used is to use spherical coordinates and expand the distribution function in Legendre polynomials

$$f(v, \theta) = \sum_{l=0}^{\infty} f^l(v) P_l(\cos \theta).$$

The Legendre polynomials have a recurrence relation

$$(n + 1)P_{n+1}(\mu) - (2n + 1)\mu P_n(\mu) + nP_{n-1}(\mu) = 0$$

which, starting from $P_0(\mu) = 1$ and $P_1(\mu) = \mu$, provides an efficient method of evaluation. Also, the orthogonality relation

$$\int_{-1}^{1} P_m(x) P_n(x) \, dx = 2\delta_{mn}/(2n + 1)$$

may be used to expand any function of $\theta$ in the assumed form. This method produces a set of coupled equations in $v$ for the coefficients of the Legendre polynomials. The infinite set has to be truncated in some way to give a finite set of equations suitable for numerical solution.

Often the full collision term is not used. Various approximations are possible in order to simplify the equations. For example, the background, used to calculate the Rosenbluth potentials in the Fokker–Planck equation, may be assumed to be isotropic or even Maxwellian.

The full equation conserves momentum and energy, so if a steady state is to be reached with a power input from waves, some kind of *ad hoc* loss term must be introduced. Alternatively, a linearised collision operator may be used, with the Rosenbluth potentials calculated with a given background. This is a good approximation if we are considering a tail of fast particles coming rapidly to a steady state under the combined effect of the waves and of collisions with the bulk of the plasma. The input of energy into the bulk can then, if necessary, be taken into account and its evolution over a longer timescale calculated from transport equations. Over this longer timescale it is, of course, necessary to update the Fokker–Planck calculation.

Simplifications of the wave diffusion term may also be used. A lot of work on lower hybrid current drive has used a simple form in which the diffusion coefficient is assumed constant between some minimum and maximum velocities and zero elsewhere. The upper limit is to be regarded as determined by the accessibility condition, as discussed in Chapter 4, and the lower end by the upper limit of the $k_{\parallel}$ spectrum generated by the antenna.

## 6.5 Lower hybrid current drive

We shall begin a more detailed discussion of the main regimes in which RF current drive is possible with an examination of the lower hybrid range of frequencies, which has proved very successful in a number of experiments. The theory of propagation and absorption of lower hybrid waves has already been given in Chapter 4, but it may be useful to include here a brief reminder of the main results of relevance to our present discussion.

For current drive to occur it is necessary for the experiment to be in the parameter range in which absorption occurs through electron Landau damping. As we have described earlier, at sufficiently high density other processes such as stochastic heating of ions or parametric decays become important, so there is a density limit above which current drive is no longer effective. The wave spectrum extends from the minimum value of $k_{\parallel}$ needed for accessibility up to a maximum value which has to be determined by the need to avoid strong absorption at the plasma edge. In general, this spectrum corresponds to a range of resonant parallel velocities from a few times thermal up to very high velocities which may be well into the relativistic range. The result is that damping takes place on the tail of the electron distribution function and a long tail on the parallel distribution function is produced, as shown in figure 6.1.

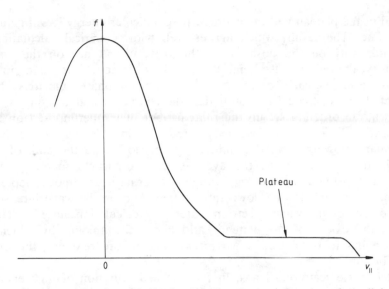

**Figure 6.1** The electron distribution function as a function of parallel velocity during lower hybrid heating. The plateau extends over the range of parallel phase velocity of the wave spectrum.

Since electrons are accelerated in the parallel direction it might be supposed that a reasonable approximation would be to use a one-dimensional approximation (Vedenov 1967, Fisch 1987), in which the distribution function is taken to be

$$f = f_0(v_\perp)f(v_\parallel) \qquad (6.18)$$

with $f_0$ a Maxwellian at the bulk plasma temperature. This leads to a simplification of the Fokker–Planck equation to an ordinary differential equation, the result of which, when compared to more exact calculations with the perpendicular degree of freedom included, gives results which are a reasonable first approximation (out by a factor of 2–3 at most).

If one looks at numerical solutions of the Fokker–Planck equation with the perpendicular degree of freedom included, it becomes apparent that the reason why the one-dimensional approximation does not give better results is that the perpendicular temperature in the region of the plateau in figure 6.1 is much larger than the bulk temperature (by a factor of 20–30 typically). The reason for this has been analysed by Fuchs *et al* (1985b) (see also Hizanidis *et al* 1985). It is the result of collisional pitch angle scattering, so that particles accelerated in the parallel direction by the waves are scattered in direction and acquire high perpendicular energies. An analytical model developed in the last mentioned papers estimates this effect by calculating from the energy and momentum slowing down equations the average time a particle will spend in the plateau and the average perpendicular energy it will gain in this time. The result, which agrees well with numerical calculations, depends only on the position of the plateau and not on the wave intensity, the only proviso being that the latter is enough to maintain an almost level plateau over the range of resonant phase velocities. The essential idea of the calculation is that the wave diffusion acts on a short timescale in order to set up the plateau, but that superimposed on the random walk in parallel velocity produced in this way there is a systematic slowing down determined by collisions with the bulk of the distribution. As a result, the average time a particle spends in the plateau, and the amount of pitch angle scattering it undergoes, depends on the collisional slowing down rate, rather than on the wave intensity. The extent of the wave spectrum enters the calculation since it determines the extent of the plateau, and hence the amount by which a particle has to be slowed down on average before rejoining the bulk distribution.

Pitch angle scattering leads, in fact, to a distribution of high energy particles all round the bulk Maxwellian, even in the opposite direction to the original wave acceleration. As described in the last section, this may lead to the production of runaway particles, in the opposite direction to the wave acceleration, if there is also a DC electric field.

The sort of numerical model needed in order to make detailed comparisons with experiment is typified by the work of Bonoli and Englade (1986) and Bonoli *et al* (1988). In this work a ray-tracing code, using the warm plasma dispersion relation, is used to follow the propagation and absorption of the wave, starting from a Brambilla spectrum at the plasma edge (see Chapter 4), calculated from the characteristics of the launching system in the machine. To do this the spectrum is divided into around 50 intervals and each of the resulting $k_\parallel$ components followed separately. The minor cross section of the plasma is divided into 40 radial zones and the quasi-linear diffusion coefficient calculated for each zone by averaging over the flux surface using the calculated wave intensity from the ray-tracing code. On each flux surface the Fokker–Planck equation with appropriate quasi-linear driving term is solved in order to calculate the current and energy deposition profiles across the minor cross section. Because of the complexity of the complete code, a one-dimensional approximation to the Fokker–Planck equation is used, but the perpendicular temperature in the plateau region is taken to be enhanced, as described above, rather than to have the bulk value. An estimate is made of the diffusion of the current resulting from cross-field transport processes and a modified profile calculated from it.

This analysis is embedded in a larger code which calculates the time evaluation of a tokamak plasma under the influence of a combination of

Ohmic and lower hybrid current drive. The radiofrequency current and energy deposition have to be calculated at each time step and used as input into a set of radial thermal transport equations and a set of equations describing the evolution of the total current and the magnetic field. The latter are of interest in the present connection, since they illustrate some of the complexities of current ramp-up using radio-frequency waves. The same considerations apply, of course, to current ramp-up using waves in any other frequency range.

Using Maxwell's equations, the evolution of the parallel electric field and the poloidal magnetic field is found to be given by

$$\partial B_\theta / \partial t = \partial E_\parallel / \partial r$$

and

$$\frac{1}{r} \frac{\partial B_\theta}{\partial r} = \mu_0 J.$$

In terms of the flux function $\psi$ defined by $\partial \psi / \partial r = B_\theta$, we obtain, by integrating the first of these with respect to $r$,

$$\partial \psi / \partial t = E_\parallel + K.$$

The constant of integration $K$ is a function of time alone and is given by the boundary condition at the edge of the plasma

$$V = -L \, dI/dt + V_{OH} + V_{EF}$$

where $V$ is the voltage around the tokamak, which depends, of course, on the parallel electric field. The first term on the right-hand side above is that contributed by the self-inductance of the plasma if the total current which it is carrying is time-dependent, the second is the loop voltage induced by changing current in the primary circuit of the Ohmic heating transformer and the third is that due to changing current in the vertical equilibrium field coils. The total current $I$ is taken to be the result of a combination of the radiofrequency current density, derived from the Fokker–Planck equation as described above and an Ohmic current density, related to the parallel electric field by a simple Ohm's law with the usual Spitzer resistivity for the plasma. Combining all these equations and assumptions, an equation for the evolution of the flux function, and hence of the poloidal magnetic field, can be obtained. Bonoli *et al* (1988) use this model to analyse the behaviour of the tokamak Alcator C, and show that the main features of lower hybrid current drive on it can be reproduced by the computer code. One prediction of this code is that the spectral gap problem, that is the fact that the parallel wavenumber spectrum generated at the plasma edge does not appear to extend to low enough velocities to pull out any significant number of particles, is resolved by the parallel wavenumber shifts which occur in the toroidal plasma. Codes like this, designed to simulate many of the features of a tokamak discharge, represent a very considerable investment of effort and require substantial computational resources to run.

Experimentally, current drive using lower hybrid waves was first observed using linear and small toroidal devices soon after the effect was predicted (Wong *et al* 1980, McWilliams and Motley 1981, Kojima *et al* 1981, La Haye *et al* 1980). In a tokamak, where a current already flows, it is more difficult to detect current drive unambiguously, since additional heating of the electrons can decrease the resistivity and so increase the current if the loop voltage remains constant. The loop voltage, which is measured at the plasma edge, can also be affected by changes in the current profile, or if the radiofrequency wave leads to an enhanced production of runaway electrons. Some of the main steps in establishing experimentally the existence of lower hybrid current drive in a tokamak were taken by Luckhardt *et al* (1982). The experiments of this group, carried out on the Versator II tokamak at Massachusetts Institute of Technology, operated in the so-called slide-away regime, in which the Ohmic field in a low density plasma pulls out a tail on the electron distribution function. The radiofrequency waves interacted with this tail to produce an enhanced current. If the wave field was applied when the Ohmic current was beginning to drop, the current could be maintained at a constant level while the loop voltage dropped through zero and settled at a small negative value. This negative value of the loop voltage ruled out an interpretation of the effect as being the result of edge heating or of runaway production. It was also shown that if the wave spectrum was launched in the opposite direction, in which there was no pre-existing tail on the distribution, then no current drive was seen. Further important advances were made by the group working with the PLT tokamak at Princeton who showed that a current could be sustained or ramped-up with lower hybrid waves (Jobes *et al* 1985). During these experiments it was shown that with input powers of several hundred kilowatts it was possible for up to 20% of the incident energy to be converted to magnetic field energy, the rate of increase of the latter being given by $\mathrm{d}(\frac{1}{2}LI^2)/\mathrm{d}t$. It was also shown that lower hybrid waves could be used to start up and sustain a tokamak, with no Ohmic discharge (Jobes *et al* 1984, Toi *et al* 1984). In these experiments the radiofrequency wave was used first to ionise the initially neutral gas then to build up the current.

Experiments at higher densities than in PLT were carried out in the Alcator C machine at MIT. This machine was notable for its high magnetic field (up to around 11 T), which allowed higher wave frequencies to be used and higher densities to be reached before energy started being deposited in the ions instead of the electrons. These experiments showed that lower hybrid current drive was effective in a higher density regime more relevant to a reactor than most of the other earlier experiments (Porkolab *et al* 1984). The densities used were in the range $10^{19}$–$10^{20}$ m$^{-3}$ and it was found that the scaling with density was in accord with the $1/n$ predicted by theory. The current was sustained at a constant value with zero loop voltage after the decay of the initial Ohmic field. At the same time 30–60% of the radiofrequency power was deposited in the bulk of the plasma, sufficient to maintain the temperature close to the initial value established in the Ohmic heating phase.

Lower hybrid current drive has now been applied to a considerable number of tokamaks and the current sustained by it has reached several megamps in the large JT-60 machine (Nagashima *et al* 1987). It is well established that the ratio of current to absorbed power is in general agreement with theory. As we have already mentioned in the chapter on lower hybrid heating, the density limit and problems of absorption near the edge cast doubts on the usefulness of lower hybrid waves in a reactor regime, and attention has been given to the fast wave in the same frequency range. However, at present there is no clear experimental evidence to show that this mode can drive current in a regime where the slow mode will not.

We have already pointed out that a number of experiments do not merely keep the current constant, but ramp it up, so that the self-inductance of the tokamak plays an important role and may produce undesirable effects such as an enhanced population of runaway electrons. Inductive effects and the link between the plasma and Ohmic heating circuits are also important in schemes for transformer recharge. The idea in such schemes is to seek to use a combination of radiofrequency and Ohmic current drive. In an initial Ohmic heating phase a current is set up in the Ohmic heating coils and as it decays it drives a current in the plasma inductively. As the current in the primary coil falls towards zero and the Ohmic current drive dies away, the radiofrequency power is switched on. With sufficient radiofrequency current drive it is possible to reverse the initial flow of energy from the primary coil to the plasma and to build up the current in the primary coil once more, the step known as transformer recharging. Once the current in the primary is built up to something approaching its initial level the whole cycle can start again. The advantage of such a scheme over steady state current drive appears if it is recognised that the plasma parameters may be varied so that each part of the cycle operates efficiently (Fisch 1982). Ohmic heating works well at high density, so that during the phase of inductive current drive the density could be increased. In a reactor the fusion power output would be greatest during this phase of the operation. Radiofrequency current drive, on the other hand, is most efficient at low density, so it is envisioned that the density would be reduced during the recharging phase. In this way the current could be kept constant with less radiofrequency power than would be necessary for steady state current drive at constant density. In a reactor, of course, this would be at the expense of pulsed power output, but would avoid the problems associated with a pulsed plasma current.

The behaviour of the tokamak during both current ramp-up and transformer recharge can be understood in terms of a simple circuit equation for the system. If we recognise that the current consists of two parts, the Ohmic current which is carried by slow collisional electrons and the RF driven current, carried by fast almost collisionless electrons, then we see that the resistive voltage around the tokamak depends on the Ohmic part of the current, while the voltage due to the self-inductance depends on the rate of change of the total current. The other contribution to the circuit equation comes from the inductive coupling to the external Ohmic heating circuit. From these considerations we obtain a circuit equation of the form

$$L \, \mathrm{d}I/\mathrm{d}t + R(I - I_{RF}) = -M \, \mathrm{d}I_{OH}/\mathrm{d}t$$

where $L$ is the self-inductance of the plasma torus, $R$ its resistance, calculated from the usual Spitzer formula, and $M$ the mutual inductance between the plasma and the Ohmic heating coil. For current ramp-up, the current in the Ohmic heating coil is kept constant, so that the right-hand side of the equation is zero and if $I_{RF} > I$ the current increases and approaches $I_{RF}$ asymptotically. For transformer recharging we again need $I_{RF} > I$, but this time $I$ is held constant and the right-hand side is non-zero. The increase in current is now opposed by an induced electric field produced by the Ohmic heating coils, and energy from the RF field is fed into these coils. If significant numbers of runaway electrons are produced by the induced electric field, then energy will be transferred to these fast particles instead. Avoiding this may put constraints on the maximum rate of transformer recharge. From these considerations it can be seen that the performance of a radiofrequency system in producing current ramp-up is very closely related to its performance in transformer recharging. The different regimes can be produced by altering feedback systems in the tokamak which can be set either to keep the plasma current constant or to keep the current in the Ohmic heating circuit constant.

### 6.6 Electron cyclotron current drive

One of the most important features of electron cyclotron, as compared with lower hybrid, current drive is the fact that the perpendicular rather than the parallel velocity is increased. Although the basic theory presented earlier predicts that the current drive efficiency will be of the same order of magnitude in the two cases, the efficiency of electron cyclotron heating is adversely affected by particle trapping in a toroidal machine. Because the magnetic field of a tokamak decreases away from the centre, the usual mirror effect can cause particles whose perpendicular velocity is sufficiently large compared to their parallel velocity to be trapped in the outer region of the torus. Clearly, electron cyclotron heating will tend to increase the likelihood of a particle being trapped, and the current drive efficiency may be less than predicted by the simplest theory (Cordey *et al* 1982). We have already outlined in a previous section the adjoint and other methods which may be used to take this effect into account. In the lower hybrid case the current is carried by electrons with high parallel velocity, and the effect of trapping is not so important.

The electrons which are driven by an electron cyclotron wave are determined by the cyclotron resonance condition

$$\omega - k_\parallel v_\parallel - n\Omega = 0 \qquad (6.19)$$

so that at different points in the absorption profile the change in $\Omega$ due to the magnetic field gradient produces excitation at different values of $v_\parallel$. In non-relativistic theory the resonant values of $v_\parallel$ are antisymmetric about the surface where $\omega = n\Omega$, and any net current depends on strong damping of the wave on one side of this surface. However, as discussed in section 5.3, relativistic effects are important and the change in $\Omega$, due to the relativistic mass shift of the electrons, produces an asymmetry. Even at comparatively low temperatures this effect has a substantial influence on the current profile (Cairns *et al* 1983). It produces an asymmetry in the profile which can produce a non-zero current, even if the wave intensity does not vary much over the resonance region. It also modifies the way the resonant velocity changes away from the resonance and produces a much broader current profile, particularly if the wave is incident from the high field side. The reason for this can be seen from equation (5.12), which shows that away from the resonance on the high field side the resonant parallel velocity varies as the square root of the distance from the resonance, while in the non-relativistic approximation it varies as the distance. The result is that we can go further from the resonance before running out of the range of parallel velocity in which there is a significant number of particles.

The relativistic shift in the resonance frequency is significant at particle energies (e.g. of the order of 1 keV) where relativistic corrections would otherwise be expected to be negligible. This is simply because of the sensitivity of the behaviour to very small shifts in the resonant velocity. At higher energies the effects of relativistic dynamics on the current drive efficiency have been investigated by Fisch (1981).

In the case of electron cyclotron heating, a decrease in current drive efficiency is produced since an increase in perpendicular velocity increases the electron mass and conservation of parallel momentum means that there is a corresponding decrease in the parallel velocity. Lower hybrid current drive may also be reduced by relativistic effects if the upper limit of the range of resonant parallel velocities is a substantial fraction of the velocity of light. Thus we can distinguish two effects of relativity which are essentially separate. First there is the shift in cyclotron frequency, resulting from the velocity dependence of the electron mass, small values of which can produce very significant effects because the electron velocities which are resonant with the wave depend very sensitively on the difference between the wave frequency and the cyclotron frequency. At higher energies relativistic effects on other aspects of the particle dynamics may be important, affecting the way in which the particle responds to the wave field. These effects are only significant if the mass shift is large enough to have a significant effect on electron dynamics other than the shift in the resonant frequency.

Current drive efficiency is greatest when particles with high parallel velocity are excited, as is apparent from equation (6.10). However, while the range of parallel velocities excited in lower hybrid heating is directly determined by the parallel spectrum, this is not the case for electron cyclotron heating. In the latter case, the range of $v_\parallel$ excited depends on the absorption profile of the wave, and for efficient current drive it is best to have absorption take place as far from the resonance as possible. This is possible if the plasma is at a high enough temperature and density, so that absorption is strong enough on the wings of the resonance. In Chapter 5 we described the main features of a scheme for electron cyclotron heating at downshifted frequencies below the electron cyclotron frequency everywhere in the tokamak. In such a scheme the absorption necessarily takes place away from the resonance and on the tail of the particle distribution function and so it is well suited to current drive.

As in the case of lower hybrid waves, attempts to model experiments closely require extensive computation. For example, in the work of O'Brien *et al* (1986) the wave diffusion term is calculated including the effects of the relativistic mass shift and of a radiofrequency beam localised in space. The general idea of the calculation is as described in section 1.4, the diffusion coefficient being related to the random walk in velocity of a particle which passes through a localised radiofrequency beam as it goes around the tokamak. Trapped and passing particles have to be treated separately since they obviously have quite different time intervals between transits through the beam. Also, the rotational transform of the field means that a passing particle will, in general, pass through the region of the beam at different poloidal positions on successive transits. This effect is taken into account in averaging the diffusion over a flux surface. A similar bounce-averaged quasi-linear term was also developed by Kerbel and McCoy (1985), initially for ions. The wave term is balanced against a Fokker–Planck term which is also bounce-averaged to take account of the trapped particles. The particle distribution function which solves the resulting equation is obtained numerically, using a finite difference scheme, and the current and absorbed power densities obtained from the appropriate moments of it. As well as predicting current drive efficiency, such a code may be used to calculate other diagnostic quantities, like the soft x-ray spectrum of a heated plasma, which also depend on the distribution function. The code of O'Brien *et al* was successful in explaining the observed x-ray spectrum from the CLEO tokamak.

The usual quasi-linear theory of the wave–particle interactions is based on the assumption that the particle receives a small velocity increment each time it passes through one of the RF beams. At high powers this may not be the case, and some analysis has been done on the non-linear dynamics of electrons (Nevins *et al* 1987, Taylor *et al* 1988), as described in Chapter 5. The influence of such non-linear effects on current drive is discussed by Kritz *et al* (1989). An interesting suggestion for obtaining effective current drive and heating is to exploit the trapping in phase space described earlier. If there is a gradient in the magnetic field or in the parallel refractive index along the orbit followed by the particle, then the region of phase space containing the trapped orbits may move adiabatically to regions of higher energy and take the particles with it. This scheme has the advantage of taking an entire group of electrons to higher energy, instead of relying on a process of diffusion.

Experimentally, electron cyclotron current drive was first observed in the Culham Levitron device in conditions of low density and temperature (density $\approx 10^{17}$ m$^{-3}$, temperature $\approx 7.5$ eV) very far from a typical tokamak or any other device with reactor potential (Start *et al* 1982). This machine contained a toroidal plasma with an annular cross section, surrounding a superconducting ring supported by magnetic forces. The wave was incident from the low field side in the extraordinary mode and, because of the small dimensions of the device, was able to tunnel through the evanescent layer between the low density cut-off and the upper hybrid resonance. At the upper hybrid resonance, partial conversion to a Bernstein mode occured and the latter then propagated inwards to be absorbed in the vicinity of the cyclotron resonance. Since the plasma was at a low temperature, it was possible to measure the current profile using probes. These showed a current density which changed sign on going through the resonance, in accordance with theory (relativistic effects being, of course, negligible in the conditions of this experiment). The current was also shown to be proportional to the radiofrequency power and to scale as $T_e/n_e$, again in agreement with theory.

Subsequent experiments have shown that electron cyclotron current drive may be produced in tokamaks, with plasmas more relevant to the fusion effort (e.g. Ando *et al* 1986, Robinson *et al* 1986). The experiments of Ando *et al*, carried out on the WT 2 tokamak, a machine with a major radius of 0.4 m, were similar to some of the early experiments on lower hybrid current drive in that they used a plasma in the low density slide away regime in which the initial Ohmic heating produces a population of accelerated particles. In this regime it was shown that the plasma current could be maintained at a constant level, with zero loop voltage or, if the power was increased, could be ramped-up while the loop voltage was driven negative. An increase in soft x-ray and electron cyclotron emission during the current drive phase suggested that the current was being carried by a suprathermal tail with a temperature of around 15 keV, while the bulk temperature remained near its initial value of 70 eV. The tail contained around 2% of the plasma electrons. If the toroidal magnetic field was increased, it was found that the current drive becomes less effective. Since the frequency of the waves was constant, increasing the magnetic field moved the position of resonance from the inside to the outside of the torus. The decrease in current was attributed to the effect of trapping when the particles were heated in the part of the flux surface where the magnetic field was smallest. In another set of experiments a plasma discharge was set up using radiofrequency heating and current drive alone. The initial heating was done with electron cyclotron waves, then lower hybrid waves were used to produce a current. If the electron cyclotron waves were switched on again later in the discharge, after a tail had been produced with the lower hybrid radiation, the rate of ramp-up was found to increase and the final current to be greater. The reason for this was that the mildly relativistic electrons accelerated in the parallel direction by the lower hybrid waves had their perpendicular velocity increased by the electron cyclotron waves and became less collisional. This synergistic interaction between the two types of heating, with a combination producing a higher current drive efficiency than either on its own, has since attracted a lot of attention and is a topic to which we shall return shortly.

A comparison of these experimental results with theory has been made by Dendy *et al* (1987), using numerical codes which can calculate the absorption of electron cyclotron waves and the emission of soft x-rays by non-Maxwellian distributions. These authors model the distribution with a superposition of a bulk Maxwellian and a tail represented by a shifted anisotropic Maxwellian, the parameters of these distributions being chosen in such a way as to give the experimental current and soft x-ray emission. A ray-tracing code is then used to follow the propagation of electron cyclotron waves into the plasma and to calculate the absorption on the assumed distribution function. From this the efficiency of current drive in the machine can be calculated, using the experimentally measured current. The conclusion is that the current drive efficiency falls well below that obtained in lower hybrid experiments, and well below that predicted from basic theory incorporated into a Fokker–Planck code. These results were based on launching of the X mode in such a direction as to assist the initial current, in which case most of the wave was predicted to be absorbed on the first pass. With launching in the opposite direction, refraction away from the plasma towards the wall of the vessel was predicted. This was consistent with the observation that in this case the current drive effect was less and almost the same as for perpendicular launching in the O mode. It is probable that in both these cases there were multiple reflections of the wave around the torus and that some part of the wave energy was eventually absorbed on the pre-existing fast tail in order to produce an enhancement of the current. The reason why the current is below that predicted theoretically (Dendy *et al* 1987) is still not understood in detail, but may be connnected with the fact that theoretical models do not, in general, include losses resulting from electron transport. Intuitively we may expect such losses to matter more for electron cyclotron than for lower hybrid current drive. In the lower hybrid case the electron is pushed immediately in the parallel direction whereas in the electron cyclotron case it is pushed in the perpendicular direction and only acquires a parallel drift after a time of the order of the collision time.

A number of other experiments have shown that the addition of electron cyclotron heating to lower hybrid current drive experiments can lead to an enhancement of the current drive efficiency (Riviere 1986). The reason for this is that the electron cyclotron waves, if in resonance with high parallel velocity particles, increase the perpendicular velocity of these particles. This reduces their collisions with the bulk and makes it easier for the lower hybrid waves to accelerate them in the parallel direction. In this way the lower hybrid tail is enhanced and the two effects reinforce each other (Krivenski *et al* 1985).

A study of the efficiency of electron cyclotron current drive in a plasma with a lower hybrid tail compared with that in a purely Maxwellian plasma with the same bulk temperature has been carried out by Fidone *et al* (1987). They show that the efficiency of current drive with electron cyclotron waves may be expected to be substantially enhanced if there is also lower hybrid heating to produce a high energy tail. The efficiency of electron cyclotron heating in a sufficiently hot

Maxwellian plasma (15 keV) was found to approach that of a cold plasma (3 keV) with a lower hybrid tail, leading the authors to suggest that a combination of the two heating schemes might be useful in the early stages of a reactor discharge, when the temperature is comparatively low, with electron cyclotron current drive alone being used at the final operating temperature.

Another beneficial effect of the enhancement of the perpendicular energy produced by electron cyclotron waves may be the suppression of the anomalous Doppler instability. This instability, discussed in section 6.8, occurs when the velocity distribution is stretched out in the parallel direction, and may take place during lower hybrid current drive.

The fact that electron cyclotron absorption may be localised around the cyclotron resonance opens the way for electron cyclotron current drive to be put to more subtle uses than simply driving the total toroidal current needed in the plasma. The fluid instabilities which produce disruptions in tokamaks are driven by magnetic field gradients around the rational surfaces in the machine. This makes it possible for a controllable current source to be used to alter the gradients in these critical regions and so enhance the stability of the discharge.

## 6.7 Minority species current drive

In the chapter on ion cyclotron heating we have seen how it may be possible for energy to be absorbed by a minority ion species when the wave is tuned to its fundamental cyclotron resonance. It is possible, at least in principle, for this effect to be exploited as a means of current drive (Fisch 1981a), in a way somewhat similar to electron cyclotron current drive.

If minority ions travelling in one direction along the field are heated (in their perpendicular degree of freedom), they collide less often with the majority species and so a relative drift between minority and majority species is set up. Any current carried by the ions is very small and effective current drive depends on the setting up of an electron drift, which can be produced if the minority and majority species are of different charge states. The mechanism is the same as that suggested for neutral beam current drive by Ohkawa (1970). It relies on the fact that the electron–ion collision frequency is proportional to $Z^2$ whereas the current carried by an ion is proportional to $Z$. Thus in a frame where the ion current vanishes, electrons tend to collide more often with the higher charge ions, acquiring a drift in the direction of motion of the latter, so that a net current results.

In order for this mechanism to work, those minority ions which are heated should have comparable collision rates with the majority ions and the electrons. If collisions with the majority ions are the dominant effect in slowing the minority ions, then the power is used to create the relative drift of the two ion species and there is little effect on the electrons. On the other hand, if electron collisions dominate then the minority ions simply lose energy to the electrons without there being any mechanism to create the anisotropy needed to produce a current. The ion velocities are so small compared to the electron thermal velocity that effects depending on the velocity dependence of collisions with an anisotropic ion distribution are negligible.

Although this may seem like a very indirect way of producing an electron current in the plasma, its efficiency is, at least in theory, comparable to the other schemes we have considered. An estimate of this efficiency can be obtained from the following arguments, based on the work of Fisch (1981a) and similar to the basic theory of electron current drive described earlier. The time-integrated momentum carried by a fast minority ion before it is absorbed into the bulk distribution is of the order of $Mv_{\parallel}/v$, where $v$ is the collision frequency. If we suppose that the wave increases the perpendicular velocity of the ion, then the ratio of the minority species momentum to the power dissipated is

$$p/P_{\mathrm{d}} = M \, \frac{\dfrac{\mathrm{d}}{\mathrm{d}v_{\perp}} \, (v_{\parallel}/v)}{\dfrac{\mathrm{d}}{\mathrm{d}v_{\perp}} \, (\tfrac{1}{2}Mv^2)} \tag{6.20}$$

in the steady state. The numerator above is, of course, non-zero because of the velocity dependence of the collision frequency.

Now we turn to the electron motion which results from this drift of the minority ions relative to the majority. We assume that

$$v = v_{\mathrm{e}} + v_{\mathrm{i}}$$

which divides the total momentum loss rate into that due to electrons and that due to ions, and also that the ions which absorb the power have velocities much greater than the ion thermal velocity, but much less than the electron thermal velocity. In this regime $v_{\mathrm{e}}$ is constant, while $v_{\mathrm{i}}$ varies as the inverse cube of the velocity, so that (6.20) gives

$$p/P_{\mathrm{d}} = \tfrac{3}{2}Mv_{\parallel}(v_{\mathrm{i}}/v)(1/vE) \tag{6.21}$$

where $E$ is the energy of the particle.

To calculate the current we shall assume that the charge state of the minority ions is $Z$, while the majority charge state is unity. Identifying the minority ions with a subscript 1 and the majority with a subscript 0 we see that since the electrons collide with the minority ions $Z^2$ times more often than with the majority, the electron drift speed in the steady state must obey the equation

$$n_1 Z^2 (v_1 - v_2) + n_0 (v_0 - v_e) = 0. \tag{6.22}$$

If it is assumed that the net momentum of the plasma is zero, then

$$n_1 M_1 v_1 + n_0 M_0 v_0 + n_e m v_e = 0. \tag{6.23}$$

This will be the case when the waves do not impart any parallel momentum to the particles, otherwise we can move to a frame of reference in which it does hold. Finally we assume that the plasma is electrically neutral, so that

$$Z n_1 + n_0 - n_e = 0. \tag{6.24}$$

The current is given by

$$J = e(n_1 Z v_1 + n_0 v_0 - n_e v_e) \tag{6.25}$$

which, combined with (6.22)–(6.24), gives

$$J = e n_1 v_1 (1 - Z)/(n_0 + n_1 Z^2)[Z n_0 + (n_e - n_0) M_1 / M_0]. \tag{6.26}$$

If we assume that the concentration of minority ions is very much less than that of the majority species, so that $n_0 \approx n_e$, this, together with (6.21), gives

$$J/P_d = \tfrac{3}{2} e v_{\parallel} (Z - Z^2) v_i / (v^2 E). \tag{6.27}$$

From this we can see that it is essential that $Z$ takes a value other than one. Also the factor $v_i/v^2$ goes to zero both as $v_i$ goes to zero and as $v_i$ becomes very large compared to $v_e$. It is greatest when $v_i$ and $v_e$ are comparable, in agreement with our earlier arguments. The current drive efficiency also depends on the parallel velocity of the resonant particles. For $Z = 2$, Fisch (1981a) shows that there is a broad maximum in the efficiency centred around a resonant parallel velocity of about five times the minority ion thermal velocity. When all the parameters are optimised in this way, it appears that the theoretical efficiency of this process is of the same order as that for lower hybrid current drive. To the best of our knowledge no experimental confirmation of this current drive scheme has been obtained at the time of writing.

## 6.8 The anomalous Doppler instability

This instability, also known as the Parail–Pogutse instability (Parail and Pogutse 1976, 1978) occurs when the electron distribution function has a long parallel tail, as is produced by lower hybrid current drive. The mechanism is most easily seen with reference to the discussion at the end of Chapter 1, where it was shown that the direction in which a particle moves in velocity space under the action of a cyclotron resonance is along a circle centred at the parallel wave phase velocity.

We consider now the situation shown in figure 6.2, where it is supposed that the tail in the distribution function is long enough that there is a significant distribution of particles at the cyclotron resonance given by

$$\omega - k_{\parallel}v_{\parallel} + \Omega = 0. \tag{6.28}$$

There will also be the possibility of resonance on the negative $v$ side where

$$\omega - k_{\parallel}v_{\parallel} - \Omega = 0$$

but we shall assume that this lies outside the non-negligible part of the particle distribution. The other possibility for wave interaction is, of course, the usual Landau damping at $v_{\parallel} \approx \omega/k_{\parallel}$. If we consider the so-called anomalous Doppler resonance given by (6.28), then it can be seen that particles diffusing under the action of the waves towards higher $v_{\perp}$ lose kinetic energy. This occurs if the tail is almost flat in the parallel direction, but falls away in the perpendicular direction, so that the distribution function will decrease away from the axis along the diffusion path. The result is that particles on average give up energy to the wave and tend to drive an instability. For the instability to occur, it is, of course, necessary for this effect to overcome Landau damping at $\omega/k_{\parallel}$. In general, it is necessary for $\omega/k_{\parallel}$ to be on the edge of the bulk distribution, where Landau damping is weak, and for there to be a long tail, reaching out to the anomalous Doppler resonance.

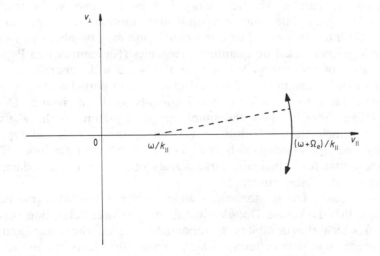

**Figure 6.2** Particle diffusion producing the anomalous Doppler instability. If the anomalous Doppler resonance lies on the plateau of the distribution function, then the distribution function will decrease away from the $v_{\parallel}$ axis, along the direction of decreasing energy on the diffusion paths.

Another way of looking at the physics of this problem is to move to the rest frame of a particle at the anomalous Doppler resonance. The Doppler shift of the wave leads to a change in the sign of its frequency (so it propagates in the opposite direction) which is essentially the origin of the term anomalous Doppler instability. In this frame of reference, a wave which was initially of positive energy becomes a negative energy wave, so an increase in the energy of the particle leads to an increase in the wave amplitude. In this frame of reference instability will occur if the distribution function decreases away from the origin along the diffusion path.

Our first picture of the wave diffusion direction allows us to say something about the energy flow in the instability. For a particle moving along the path shown, the ratio of the change in parallel to the change in perpendicular energy is

$$\frac{2v_\parallel \Delta v_\parallel}{(\Delta v_\perp)^2}$$

which can be seen from simple geometry to be (for small $\Delta v_\parallel$, $\Delta v_\perp$) approximately

$$\frac{v_\parallel k_\parallel}{\Omega} = \frac{\omega + \Omega}{\Omega}.$$

Thus, of the parallel kinetic energy lost by the particles a fraction $\Omega/(\omega + \Omega)$ goes into the perpendicular motion, and a fraction $\omega/(\omega + \Omega)$ into the wave. The same conclusion may be obtained from a simple argument based on quantum mechanics (Kadomtsev and Pogutse 1968). Part of the energy going into the wave will generally be lost through Landau damping to the parallel motion of particles at $v_\parallel = \omega/k_\parallel$. A more detailed examination of this instability has been given by Dendy *et al* (1986), showing how the simple physical picture we have given, based on single particle orbits, is related to the more usual type of stability analysis working with the plasma distribution function. They also show that for sufficiently large values of $k_\parallel$, ion Landau damping may have a stabilising effect.

In low density Ohmic discharges where runaway electrons produce a long tail, the anomalous Doppler instability produces relaxation oscillations in which the instability is repeatedly excited then stabilised by quasi-linear diffusion effects, which cause the distribution to be broadened in the perpendicular velocity direction (Knoepfel and Spong 1979). In the presence of lower hybrid current drive, the relaxation oscillations are lengthened in period and may even be suppressed (Luckhardt *et al* 1986). However, even in the absence of relaxation oscillations it is possible that the anomalous Doppler effect plays a role in enhancing the perpendicular temperature in the lower hybrid tail, adding to the effects of collisional pitch angle scattering.

Luckhardt *et al* (1987) have studied the anomalous Doppler instability for realistic lower hybrid current drive electron distributions. Using the analytic model of Fuchs *et al* (1985b) for the perpendicular temperature in the tail, they obtain analytic estimates of the growth rate. For there to be instability it is necessary that the anomalous Doppler resonance and the Landau resonance both lie on the plateau of the electron distribution. The normal cyclotron resonance at $(\omega - \Omega_e)/k_\parallel$, which also contributes some damping, must be beyond the bulk of the distribution on the opposite side from the plateau. Using these conditions, Luckhardt *et al* map out the frequency and parallel wavenumber ranges in which instability might be expected for a number of different machines.

## 6.9 Conclusion

In keeping with the theme of this book, the discussion has been confined to radiofrequency methods of current drive, though other ideas have been proposed and are discussed by Fisch (1987). At present, lower hybrid waves appear to be capable of driving the current required to sustain a tokamak discharge, while electron cyclotron current drive has been shown to exist but so far at a disappointingly low efficiency.

In extrapolating to a reactor regime some doubts appear, because current drive efficiency generally decreases with density and the ability of lower hybrid waves to penetrate to the centre of the plasma is in doubt. As discussed in the chapter on lower hybrid heating, the fast mode provides a possible way around the problems associated with the usual slow mode lower hybrid heating, but experimental evidence of its advantages is still lacking.

The efficiency of all current drive schemes found so far is lower than that of Ohmic heating and a somewhat higher efficiency would be desirable in a reactor in order to reduce the recirculating power. However, even if it proves uneconomic to drive current continuously by this means, it is still likely that radiofrequency current drive could play a useful role in modifying the current profile in order to suppress instabilities.

# References

Abe H, Okada H, Itatani R, Ono M and Okuda H 1984 *Phys. Rev. Lett.* **53** 1153

Adam J 1984 *Plasma Phys. Control. Fusion* **26** 165

Airoldi A C and Orefice A 1982 *J. Plasma Phys.* **27** 515

Akhiezer A I, Akhiezer I A, Polovin R V, Sitenko A G and Stepanov K N 1975 *Plasma Electrodynamics* (Oxford: Pergamon)

Alikaev V V, Bobrovskij G A, Poznyak V I, Razumova K A, Sannikov V V, Sokolov Yu A and Shmarin A A 1976 *Sov. J. Plasma Phys.* **2** 212

Alikaev V V and Parail V V 1984 *Proc. 4th Int. Symp. on Heating in Toroidal Plasmas* (Varenna: International School of Plasma Physics)

Allis W P, Buchsbaum S J and Bers A 1963 *Waves in Anisotropic Plasmas* (Cambridge: MIT Press)

Anderson D, Core W, Eriksson L-G, Hamnen H, Hellsten T and Lisak M 1987 *Nucl. Fusion* **27** 911

Ando A *et al* 1986 *Phys. Rev. Lett.* **56** 2180

Antonsen T M and Chu K R 1982 *Phys. Fluids* **25** 1295

Antonsen T M and Manheimer W M 1978 *Phys. Fluids* **21** 2295

—— 1980 *Phys. Fluids* **23** 660

Appert K, Balet B, Gruber R, Troyon F, Tsunematsu T and Vaclavik J 1982 *Nucl. Fusion* **22** 903

Appert K, Collins G A, Hellsten T, Vaclavik J and Villard L 1986 *Plasma Phys. Control. Fusion* **28** 133

Appert K and Vaclavik J 1983 *Plasma Phys.* **25** 551

Appert K, Vaclavik J and Villard L 1984 *Phys. Fluids* **27** 432

Barston E M 1964 *Ann. Phys., NY* **29** 282

Behn R, Collins G A, Lister J B and Weisen H 1987 *Plasma Phys. Control. Fusion* **29** 75

Bellan P M and Porkolab M 1974 *Phys. Fluids* **17** 1592

—— 1976 *Phys. Fluids* **19** 995

Berger J M, Newcomb W A, Dawson J M, Frieman E A, Kulsrud R M and Lenard A 1958 *Phys. Fluids* **1** 301

Berk H L and Book D L 1969 *Phys. Fluids* **12** 649

Berk H L and Dominguez R R 1977 *J. Plasma Phys.* **18** 31

Bernabei S, Heald M A, Hooke W M and Paoloni F J 1975 *Phys. Rev. Lett.* **34** 866
Bernabei S and Motley R W 1987 *Applications of Radio Frequency Power to Plasmas* (AIP Conf. Proc. 159)
Bernabei S *et al* 1982 *Phys. Rev. Lett.* **49** 1255
Bernstein I B and Baxter D C 1981 *Phys. Fluids* **24** 108
Bernstein I B and Friedland L 1983 in *Handbook of Plasma Physics* vol. 1 ed. M N Rosenbluth and R Z Sagdeev (Amsterdam: North-Holland)
Bers A 1975 in *Plasma Physics, Les Houches 1972* ed. C De Witt and J Peyraud (New York: Gordon and Breach)
—— 1976 in *Proc. 3rd Symp. on Heating in Toroidal Devices, Varenna* (Bologna: Editrice Compositori)
—— 1978 in *Proc. 3rd Topical Conf. on R.F. Plasma Heating* (Pasadena: Calif. Inst. of Tech.) pp A1-1–A1-10
Bers A (ed.) 1984 *IEEE Trans. Plasma Sci.* **PS-12** no 2 (Special Issue on RF Heating and Current Generation)
Bers A, Jacquinot J and Lister G 1980 *Heating in Toroidal Plasmas, Proc. 2nd Joint Grenoble–Varenna Int. Symp.* (Brussels: Commission of the European Communities)
Bers A and Theilhaber K 1983 *Nucl. Fusion* **23** 41
Besson G, de Chambrier A, Collins G A, Joye B, Lietti A, Lister J B, Moret J M, Nowak S, Simm S and Weisen H 1986 *Plasma Phys. Control. Fusion* **28** 1291
Bhatnagar V P, Koch R, Messiaen A M and Weynants R 1982 *Nucl. Fusion* **22** 280
Bickerton R J 1972 *Comm. Plasma Phys. Control. Fusion* **1** 95
Bonoli P T and Englade R C 1986 *Phys. Fluids* **29** 2937
Bonoli P T and Ott E 1982 *Phys. Fluids* **25** 359
Bonoli P T, Porkolab M, Takase Y and Knowlton S F 1988 *Nucl. Fusion* **28** 991
Borg G G, Dalla Piazza S, Martin Y, Pochelon A, Ryter F and Weisen H 1989 *Proc. 16th European Conf. on Controlled Fusion and Plasma Phys.* (Geneva: European Physical Society) p 1199
Bornatici M, Cano R, De Barbieri O and Engelmann F 1983 *Nucl. Fusion* **23** 1153
Bornatici M and Engelmann F 1980 *Phys. Fluids* **23** 659
Bornatici M, Engelmann F, Maroli C and Petrillo V 1981 *Plasma Phys.* **23** 89
Bornatici M and Ruffina U 1986 *Plasma Phys. Control. Fusion* **28** 1589
Brambilla M 1976 *Nucl. Fusion* **16** 47
—— 1979 *Nucl. Fusion* **19** 1343
—— 1982 *Heating in Toroidal Plasmas, Proc. 3rd Joint Grenoble–Varenna Int. Symp.* (Brussels: Commission of the European Communities)
—— 1986 *Comp. Phys. Reps.* **4** 73
Brambilla M and Krucken T 1988 *Nucl. Fusion* **28** 1813
Buchsbaum S J 1960 *Phys. Fluids* **3** 418
Budden K G 1961 *The Propagation of Radio Waves in the Ionosphere* (Cambridge: Cambridge University Press)
—— 1985 *The Propagation of Radio Waves* (Cambridge: Cambridge University Press)

Cairns R A and Fuchs V 1989 *Phys. Fluids* B **1** 350
Cairns R A and Lashmore-Davies C N 1982 *Phys. Fluids* **25** 1602
—— 1983 *Phys. Fluids* **26** 1268
—— 1986a *Phys. Fluids* **29** 3639
—— 1986b *Plasma Phys. Control. Fusion* **28** 1539
Cairns R A, Owen J and Lashmore-Davies C N 1983 *Phys. Fluids* **26** 3475
Canobbio E and Croci E 1984 *Proc. 4th Int. Symp. on Heating in Toroidal Plasmas* (Varenna: International School of Plasma Physics)
Case K M 1959 *Ann. Phys., NY* **7** 349
Cavallo A, Hsuan H, Boyd D, Grek B, Johnson D, Kritz A, Mikkelsen D, LeBlanc B and Takahashi H 1985 *Nucl. Fusion* **25** 335
Chan V and Guest G 1982 *Nucl. Fusion* **22** 272
Chandrasekhar S 1943 *Rev. Mod. Phys.* **15** 1
Chen L and Hasegawa A 1974 *Phys. Fluids* **17** 1399
Chu K R 1985 in *Wave Heating and Current Drive in Plasmas* ed. V L Granatstein and P L Colestock (New York: Gordon and Breach)
Chu K R and Hui B 1983 *Phys. Fluids* **26** 69
Colborn J A, Parker R R, Chen K-I, Luckhardt S C and Porkolab M 1987 in *Applications of Radio Frequency Power to Plasmas* ed. S Bernabei and R W Motley (New York: American Institute of Physics)
Colestock P L 1985 in *Wave Heating and Current Drive in Plasmas* ed. V L Granatstein and P L Colestock (New York: Gordon and Breach)
Colestock P L and Kashuba R J 1983 *Nucl. Fusion* **23** 763
Colestock P L and Kulp J L 1980 *IEEE Trans. Plasma Sci.* **PS-8** 71
Collins G A, Hofmann F, Joye B, Keller R, Lietti A, Lister J B and Pochelon A 1986 *Phys. Fluids* **29** 2260
Cordey J G, Edlington T and Start D F H 1982 *Plasma Phys.* **24** 73
Cross R 1988 *Alfven Waves* (Bristol: Adam Hilger)

Demeio L and Engelmann F 1986 *Plasma Phys. Control. Fusion* **28** 1851
Dendy R O, Lashmore-Davies C N and Montes A 1986 *Phys. Fluids* **29** 4040
Dendy R O, O'Brien M R, Cox M and Start D F H 1987 *Nucl. Fusion* **27** 377
Dnestrovskii Yu N and Kostomarov D P 1961 *Sov. Phys.–JETP* **13** 986
Dreicer H 1960 *Phys. Rev.* **117** 329

Ejima S and Prater R 1987 *Nucl. Fusion* **27** 1135
Eldridge O C, England A C, Gilgenbach R M, Hackett K F, Kulcher A G, Loring C M and Wilgen J B 1980 *Proc. 2nd Joint Grenoble–Varenna Int. Symp. on Heating in Toroidal Plasmas* vol. 1 (Brussels: Commission of the European Communities) p 99

Ferreira A, O'Brien M R and Start D F H 1984 *Plasma Phys. Control. Fusion* **26** 1565
Fidone I, Giruzzi G, Krivenski V, Mazzucato E and Ziebell L F 1987 *Nucl. Fusion* **27** 579
Fidone I, Giruzzi G, Krivenski V and Ziebell L F 1986 *Phys. Fluids* **29** 803
Fidone I, Giruzzi G and Mazzucato E 1985 *Phys. Fluids* **28** 1224

Fidone I, Granata G and Meyer R L 1982 *Phys. Fluids* **25** 2249
Fidone I, Granata G, Ramponi G and Meyer R L 1978 *Phys. Fluids* **21** 645
Fisch N J 1978 *Phys. Rev. Lett.* **41** 873
—— 1981 *Phys. Rev. Lett.* **24** 3245
—— 1981a *Nucl. Fusion* **21** 15
—— 1982 *Heating in Toroidal Plasmas, Proc. 3rd Joint Varenna–Grenoble Int. Symp.* vol. 3 (Brussels: Commission of the European Communities) p 841
—— 1987 *Rev. Mod. Phys.* **59** 175
Fisch N J and Boozer A H 1980 *Phys. Rev. Lett.* **45** 720
Fisch N J and Karney C F F 1985 *Phys. Fluids* **28** 3107
—— 1985a *Phys. Rev. Lett.* **54** 897
Friedland L 1985 *Phys. Fluids* **28** 3260
—— 1986 *Phys. Fluids* **29** 1105
Friedland L and Goldner G 1986 *Phys. Fluids* **29** 4073
Friedland L, Goldner G and Kaufman A N 1987 *Phys. Rev. Lett.* **58** 1392
Friedland L and Kaufman A N 1987 *Phys. Fluids* **30** 3050
Fuchs V and Bers A 1988 *Phys. Fluids* **31** 3702
Fuchs V, Bers A and Harten L 1985a *Phys. Fluids* **28** 177
Fuchs V, Cairns R A, Shoucri M M, Hizanidis K and Bers A 1985b *Phys. Fluids* **28** 3619
Fuchs V, Ko K and Bers A 1981 *Phys. Fluids* **24** 1251
Furth H P, Glasser A H, Park W, Rutherford P, Selberg H and White R B 1985 *Proc. 12th European Conf. on Plasma Phys. and Controlled Fusion* (Geneva: European Physical Society)

Gambier D J D and Schmitt J P M 1983 *Phys. Fluids* **26** 2200
Golant V E 1972 *Sov. Phys.–Tech. Phys.* **16** 1980
Goldstein H 1980 *Classical Mechanics* (2nd edn) (Reading, Massachusetts: Addison Wesley)
Granatstein V L and Colestock P L (ed.) 1985 *Wave Heating and Current Drive in Tokamaks* (New York: Gordon and Breach)

Hasegawa A and Uberoi C 1982 *The Alfven Wave* (Springfield VA: Technical Information Center, US Dept of Energy)
Hasegawa A and Chen L 1976 *Phys. Fluids* **19** 1924
Hawryluk R J and Schmidt J A 1976 *Nucl. Fusion* **16** 775
Heading J 1961 *J. Res. Natl. Bur. Stand.* D **65** 595
Hellsten T, Appert K, Vaklavik J and Villard L 1985 *Nucl. Fusion* **25** 99
Hellsten T and Villard L 1988 *Nucl. Fusion* **28** 285
Hinton F and Hazeltine R D 1976 *Rev. Mod. Phys.* **42** 239
Hizanidis K, Bers A, Fuchs V and Cairns R A 1985 *Phys. Fluids* **29** 1331
Hofmann F, Appert K and Villard L 1984 *Nucl. Fusion* **24** 1679
Hosea J *et al* 1985 *Proc. 12th European Conf. on Controlled Fusion and Plasma Phys.* (Geneva: European Physical Society)

Ignat D W 1981 *Phys. Fluids* **24** 1110
Itoh K, Itoh S-I and Fukuyama A 1984 *Nucl. Fusion* **24** 13

Jacquinot J, McVey B D and Scharer J E 1977 *Phys. Rev. Lett.* **39** 88
Jacquinot J, Theilhaber K, Lister G and Brambilla M 1982 *Heating in Toroidal Plasmas, Proc. 3rd Joint Varenna–Grenoble Int. Symp.* (Brussels: Commission of the European Communities)
Jacquinot J *et al* 1986 *Plasma Phys. Control. Fusion* **28** 1
Jacquinot J *et al* 1987 in *Applications of Radio Frequency Power to Plasmas* ed. S Bernabei and R W Motley (New York: American Institute of Physics)
Jaeger E F, Batchelor D B and Weitzner H 1988 *Nucl. Fusion* **28** 53
Jaeger E F, Batchelor D B, Weitzner H and Whealton J H 1986 *Comput. Phys. Comm.* **40** 33
Jobes F C, Bernabei S, Chu T K, Hooke W M, Meservey E B, Motley R W, Stevens J E and Von Goeler S 1985 *Phys. Rev. Lett.* **55** 1295
Jobes F C *et al* 1984 *Phys. Rev. Lett.* **52** 1005

Kadomtsev B B and Pogutse O P 1968 *Sov. Phys.–JETP* **26** 1146
Karney C F F 1979 *Phys. Fluids* **21** 1591
—— 1986 *Comp. Phys. Reps.* **4** 183
Karney C F F and Bers A 1977 *Phys. Fluids* **39** 550
Karney C F F and Fisch N J 1981 *Nucl. Fusion* **21** 1549
—— 1986 *Phys. Fluids* **29** 180
Kay A, Cairns R A and Lashmore-Davies C N 1986 *Proc. 18th European Conf. on Plasma Phys. and Controlled Fusion* vol. 2 (Geneva: European Physical Society) p 93
—— 1988 *Plasma Phys. Control. Fusion* **30** 471
Kaye S M and Goldston R J 1985 *Nucl. Fusion* **25** 65
Keilhacker M *et al* 1984 *Plasma Phys. Control. Fusion* **26** 49
Kennel C F and Engelmann F 1966 *Phys. Fluids* **9** 2377
Kerbel G D and McCoy M G 1985 *Phys. Fluids* **28** 3629
Knoepfel H and Spong D 1979 *Nucl. Fusion* **19** 785
Koch R, Bhatnagar V B, Messiaen A M and van Ester D 1986 *Comput. Phys. Comm.* **40** 1
Kojima T, Takamura S and Okuda T 1981 *Phys. Lett.* **83A** 172
Kreischer K E, Danly B G, Saito H, Schutkeker J B, Temkin R E and Tran T M 1985 *Plasma Phys. Control. Fusion* **27** 1449
Kritz A H, Smith G R, Nevins W M and Cohen R H 1989 *Phys. Fluids* B **1** 142
Krivenski V and Orefice A 1983 *J. Plasma Phys.* **30** 125
Krivenski V, Fidone I, Giruzzi G, Granata G, Meyer R L and Mazzucato E 1985 *Nucl. Fusion* **25** 127

La Haye R J, Armentrout C J, Harvey R W, Moeller C P and Stambaugh R D 1980 *Nucl. Fusion* **20** 218
Lallia P 1974 in *Proc. 2nd Topical Conf. on Radio Frequency Plasma Heating, Lubbock, Texas*
Landau L D 1936 *Phys. Z. Sowjetunion* **10** 154
Lashmore-Davies C N 1972 *Plasma Phys.* **14** 357
Lashmore-Davies C N, Cairns R A and Fuchs V 1985 *Phys. Fluids* **28** 1791
Lashmore-Davies C N, Fuchs V, Francis G, Ram A K, Bers A and Gauthier L 1988 *Phys. Fluids* **31** 3702

Lazzaro E and Ramponi G 1981 *Plasma Phys.* **23** 53
Leclert G P, Karney C F F, Bers A and Kaup D J 1979 *Phys. Fluids* **22** 1545
Littlejohn R G 1983 *J. Plasma Phys.* **15** 125
—— 1985 *Phys. Fluids* **28** 2015
Litvak A G, Permitin G V, Suvorov E V and Frajman A A 1977 *Nucl. Fusion* **17** 659
Luckhardt S C, Bers A, Fuchs V and Shoucri M 1987 *Phys. Fluids* **30** 2110
Luckhardt S C, Chen K-I, Mayberry M J, Porkolab M, Terumichi Y, Bekefi G, McDermott F S and Rohatgi R 1986 *Phys. Fluids* **29** 1985
Luckhardt S C, Porkolab M, Knowlton S F, Chen K-I, Fisher A S, McDermott F S and Mayberry M 1982 *Phys. Rev. Lett.* **48** 152

MacKay R S and Meiss J D 1987 *Hamiltonian Dynamical Systems* (Bristol: Adam Hilger)
Mahajan S M 1984 *Phys. Fluids* **27** 2238
Matsuda K 1986 *Phys. Fluids* **29** 2493
McWilliams R and Motley R W 1981 *Phys. Fluids* **24** 2022
Mett R R and Tataronis J A 1989 *Phys. Rev. Lett.* **63** 1380
Mjølhus E 1987 *J. Plasma Phys.* **38** 1
Morales G J and Lee Y C 1975 *Phys. Rev. Lett.* **35** 930
Morishita T, Fukuyama A, Hamamatsu K, Itoh S-I and Itoh K 1987 *Nucl. Fusion* **27** 1291

Nagashima T *et al* 1987 in *Applications of Radio Frequency Power to Plasmas* ed. S Bernabei and R W Motley (New York: American Institute of Physics)
Nevins W M, Rognlien T D and Cohen B I 1987 *Phys. Rev. Lett.* **59** 60
Ngan Y C and Swanson D G 1977 *Phys. Fluids* **20** 1920

O'Brien M R, Cox M and Start D F H 1986 *Nucl. Fusion* **26** 1625
Ohkawa T 1970 *Nucl. Fusion* **10** 185
Ohkawa T, Chan V S, Chu M S, Dominguez R R and Miller R L 1988 *Proc. Twelfth Int. Conf. on Plasma Phys. and Controlled Nuclear Fusion* (Vienna: International Atomic Energy Agency) Paper CN50/I-4
Ohkubo K, Hamada Y and Ogawa Y 1986 *Phys. Rev. Lett.* **56** 2040
Ono M 1980 *Phys. Rev. Lett.* **45** 1105
—— 1982 *Phys. Fluids* **25** 990
—— 1987 in *Applications of Radio Frequency Power to Plasmas* ed. S Bernabei and R W Motley (New York: American Institute of Physics)
Ono M, Wong K L and Wurden G A 1983 *Phys. Fluids* **26** 298
Orefice A 1988 *J. Plasma Phys.* **39** 61

Parail V V and Pogutse O P 1976 *Sov. J. Plasma Phys.* **2** 126
—— 1978 *Nucl. Fusion* **18** 303
Peng Y K-M, Borowski S K and Kammash T 1978 *Nucl. Fusion* **18** 1489
Perkins F W 1977 *Nucl. Fusion* **17** 1197
Pert G J 1978 *Plasma Phys.* **20** 175
Pinsker R I, Colestock P L, Bernabei S, Cavallo A, Greene G J, Kaita R and Stevens J E 1987 in *Applications of Radio Frequency Power to Plasmas* ed. S Bernabei and R W Motley (New York: American Institute of Physics)

Pinsker R I, Duvall R E, Fortgang C M and Colestock P L 1986 *Nucl. Fusion* **26** 941

Porkolab M 1977 *Phys. Fluids* **20** 2058

—— 1985a *Phys. Rev. Lett.* **54** 434

—— 1985b in *Wave Heating and Current Drive in Plasmas* ed. V L Granatstein and P L Colestock (New York: Gordon and Breach)

Porkolab M *et al* 1984 *Phys. Rev. Lett.* **53** 450

Prater R, Ejima S, Harvey R W, Lieber A J, Matsuda K and Moeller C 1986 *Proc. 11th Int. Conf. on Plasma Phys. and Controlled Fusion Research* (Vienna: International Atomic Energy Agency)

Preinhalter J and Kopeky V 1973 *J. Plasma Phys.* **10** 1

Puri S 1979 *Phys. Fluids* **22** 1716

—— 1987 *Nucl. Fusion* **27** 229


Ram A K and Bers A 1984 *Nucl. Fusion* **24** 697

—— 1987 in *Applications of Radio Frequency Power to Plasmas* ed. S Bernabei and R W Motley (New York: American Institute of Physics)

Read M E and Granatstein V L 1985 in *Wave Heating and Current Drive in Plasmas* ed. V L Granatstein and P L Colestock (New York: Gordon and Breach)

Riviere A C 1986 *Plasma Phys. Control. Fusion* **28** 1263

—— 1987 in *Applications of Radio Frequency Power to Plasmas* ed. S Bernabei and R W Motley (New York: American Institute of Physics)

Robinson D C *et al* 1986 *Proc. Int. Conf. on Plasma Phys. and Controlled Fusion* (Vienna: International Atomic Energy Agency)

Robinson P A 1986 *J. Math. Phys.* **27** 1206

—— 1987 *J. Math. Phys.* **28** 1203

Romero H and Scharer J 1987 *Nucl. Fusion* **27** 363

Rosenbluth M N, Coppi B and Sudan R N 1969 *Ann. Phys., NY* **55** 248

Rosenbluth M N, MacDonald W M and Judd D L 1957 *Phys. Rev.* **107** 1

Ross D W, Yanming Li, Mahajan S M and Michie R B 1986 *Nucl. Fusion* **26** 139


Scharer J E, McVey B D and Mau T K 1977 *Nucl. Fusion* **17** 297

Sedláček Z 1971 *J. Plasma Phys.* **5** 239; **6** 187

Shkarofsky I P 1966a *Phys. Fluids* **9** 561

—— 1966b *Phys. Fluids* **9** 570

Shohet J 1978 *Comm. Plasma Phys. Control. Fusion* **4** 37

Smithe D N, Colestock P L, Kashuba R J and Kammash T 1987 *Nucl. Fusion* **27** 1319

Start D F H, Ainsworth N R, Cordey J G, Edlington T, Fletcher W H W, Payne M F and Todd T N 1982 *Phys. Rev. Lett.* **48** 624

Start D F H *et al* 1987 in *Applications of Radio Frequency Power to Plasmas* ed. S Bernabei and R W Motley (New York: American Institute of Physics)

Steinmetz K 1987 in *Applications of Radio Frequency Power to Plasmas* ed. S Bernabei and R W Motley (New York: American Institute of Physics)

Stix T H 1962 *The Theory of Plasma Waves* (New York: McGraw Hill)

—— 1965 *Phys. Rev. Lett.* **15** 878

—— 1975 *Nucl. Fusion* **15** 737
Suvarov E V and Tokman M D 1983 *Plasma Phys.* **25** 723
Swanson D G 1980 *Nucl. Fusion* **20** 949
—— 1981 *Phys. Fluids* **24** 2035
—— 1985 *Phys. Fluids* **28** 2645
—— 1989 *Plasma Waves* (Boston: Academic Press)

Taguchi M 1983 *J. Phys. Soc. Japan* **52** 2035
Tataronis J and Grossman W 1973 *Z. Phys.* **261** 203
Taylor A W, Cairns R A and O'Brien M R 1988 *Plasma Phys. Control. Fusion* **30** 1039
Taylor J B 1989 *Phys. Rev. Lett.* **63** 1384
Theilhaber K 1982 *Nucl. Fusion* **22** 363
—— 1984 *Nucl. Fusion* **24** 1383
Theilhaber K and Bers A 1980 *Nucl. Fusion* **20** 547
Theilhaber K and Jacquinot J 1984 *Nucl. Fusion* **24** 541
Thomassen K I 1988 *Plasma Phys. Control. Fusion* **30** 57
Toi K *et al* 1984 *Phys. Rev. Lett.* **52** 2144
Tripathi V K, Liu C S and Chiu S C 1987 *Nucl. Fusion* **27** 287
Trubnikov B A 1959 in *Plasma Physics and the Problem of Thermonuclear Reactions* vol. 3 ed. M A Leontovich (London: Pergamon)
Tubbing B J D, Jacquinot J J, Stork D and Tanga A 1989 *Nucl. Fusion* **29** 1953

Uesugi Y *et al* 1987 in *Applications of Radio Frequency Power to Plasmas* ed. S Bernabei and R W Motley (New York: American Institute of Physics)

Van Kampen N G 1955 *Physica* **21** 949
Vedenov A A 1967 in *Reviews of Plasma Physics* vol. 3 ed. M A Leontovich (New York: Consultants Bureau)
Villalon E and Bers A 1980 *Nucl. Fusion* **20** 243
Villard L, Appert K, Gruber R and Vaclavik J 1986 *Comp. Phys. Rep.* **4** 95

Wegrowe J-G and Tonon G 1983 in *Non-Inductive Current Drive in Tokamaks, Proc. IEAE Technical Committee, Culham Laboratory*
Weinberg S 1962 *Phys. Rev.* **126** 1899
Weitzner H and Batchelor D B 1979 *Phys. Fluids* **22** 1355
Wersinger J M, Ott E and Finn J M 1982 *Phys. Fluids* **25** 359
Wesson J 1987 *Tokamaks* (Oxford: Oxford University Press)
Weynants R R, Messiaen A M, Leblud C and Vandenplas P E 1980 *Heating in Toroidal Plasmas, Proc. 2nd Joint Grenoble–Varenna Int. Symp.* (Brussels: Commission of the European Communities)
Wong K-L, Horton R and Ono M 1980 *Phys. Rev. Lett.* **45** 117
Woods A M, Cairns R A and Lashmore-Davies C N 1986 *Phys. Fluids* **29** 3719
Wort D J M 1971 *Plasma Phys.* **13** 258

Ye H and Kaufman A N 1988 *Phys. Rev. Lett.* **61** 2762

# Index

accessibility condition for lower hybrid
waves, 72–6
accumulation point of Alfven wave
spectrum, 35
adjoint methods, 127–30
Airy function, 90
Alcator C, 135, 136
Alfven speed, 27
Alfven waves, 2, 23–9, 122
angular dependence of electron
cyclotron absorption, 107–8
anomalous Doppler instability, 146–9
antenna,
Alfven wave, 37–9
fast wave, 93–5
ion cyclotron, 56–7, 61–4
lower hybrid, 89–90
antenna impedence, 38
Appleton–Hartree dispersion relation,
98
ASDEX, 63

Bernstein modes, 5, 11, 47, 49, 68,
103, 109–10, 141
Bernstein wave heating, 67–70
beryllium, in JET, 42, 65, 67
Brambilla spectrum, 88–9
Budden's equation, 8, 53

Case–Van Kampen modes, 52
chaotic orbits, 81
circuit equation for tokamak, 138
circularly polarised components, 20, 44
CLEO, 141

CMA diagram, 100
cold plasma approximation, 5
confinement scaling laws, 65–6
congruent reduction, 17
continuous spectrum, 31, 40
current drive efficiency, 121–4, 143
current feeders, 57
cut-off, 48
cyclotron damping, 47
cyclotron frequency, 4

density fluctuations, 85
density limit for lower hybrid current
drive, 87, 95, 132
density limits for electron cyclotron
heating, 99–101
dielectric filled waveguides, 94
dielectric tensor, 4–5
diffusion coefficient, 19, 23
diffusion in velocity space, 17–19
diffusion paths, 23–4
dispersion relation, 5
disruptions, 119
Doublet III, 117, 118
downshifted frequencies for electron
cyclotron heating, 110, 140

edge region, ion cyclotron propagation
in, 57–9
effective field for electron cyclotron
heating, 21
eikonal approximation, 7–9
electron cyclotron current drive,
138–44

electron heating by ion cyclotron
    waves, 64
electron inertia, 30, 44
energy confinement, 65–7, 117
ergodic ray paths, 78
extraordinary (X) mode, 99

Faraday screen, 56, 61
fast magnetosonic wave, 11, 28–9, 35,
    44
fast wave heating, 93–6
filamentation, 84
flux surface, 19
focusing of ion cyclotron waves, 45
Fokker–Planck equation, 18, 59–60,
    84, 125, 127, 129–32, 134
free electron laser, 97

global modes, 36
Green's function, 50–1
grill for lower hybrid wave launch,
    87–8
gyrotron, 3, 115–16

H mode, 63
Hall effect, 34–5
Hamilton's equations, 7, 78, 112
heating efficiency, 64
helicity, 122
hot plasma mode (lower hybrid),
    79–80
hybrid resonances, 44

impurities, 67
incremental confinement time, 65
inverse Fourier transform, 11
ionisation by a radiofrequency wave,
    118–19, 136
ISX-B, 116

JET (Joint European Torus), 3, 42,
    62–3, 65, 66–7
JT-60, 137

kinetic Alfven wave, 32–3

L mode, 63
Lagrangian, 111–12

Landau damping, 25, 28, 31, 51, 92,
    95, 121
    of lower hybrid waves, 71, 83, 132
Langevin equations, 124–7
Laplace's method, 11–12
Larmor radius, 11, 12
Legendre polynomial expansion, 60,
    131
Levitron, 141–2
linear turning point, 79–80
loop antenna, 94
loop voltage, 118, 135–7
loss-cone distribution, 107
low frequency current drive, 122
lower hybrid current drive, 132–8
lower hybrid frequency, 67, 71
lower hybrid resonance, 44, 45, 57, 58,
    68, 71, 73, 79

magnetic field energy, 136
magnetosonic waves, 26–7
minority heating, 52–5, 60, 64
minority species current drive, 144–6
mode conversion, 9–17, 47–52, 74–6,
    109, 141
    between O and X modes, 101–2
    scheme for ion cyclotron heating, 53
monster sawteeth, 66
MTX, 111

non-linear electron cyclotron heating,
    111–15, 141
non-linear Landau damping, 68

Ohmic heating, 1
Ohm's law, 27, 34
ordinary (O) mode, 99

parabolic cylinder function, 14
Parail–Pogutse instability, 146–9
parametric decay
    of Alfven wave, 33
    of lower hybrid wave, 86–7, 132
parasitic interaction, 41, 61
particle motion in a wave, 19–24
perpendicular temperature with lower
    hybrid current drive, 133–4
phase space buckets, 115

phase space Lagrangian, 111
pitch angle scattering, 126, 134
PLT, 70, 136
ponderomotive force, 84–5
profile consistency, 117
profile control, 2, 98, 119, 144

quasi-linear theory, 19
quasi-modes, 86

radiation barrier, 118
ramp-up of current, 135–7
random walk, 18, 22
ray tracing, 7–9, 55, 77
reactor, 1, 118, 124
reflection coefficient, 15
refractive index, 4
relativistic dielectric tensor, 105–7
relativistic effects, 98, 102–11, 139–40
relativistic Vlasov equation, 105
relaxation oscillation, 148
resonance, 8
resonance cones, 76–8, 84, 90
resonant absorption, 33
Rosenbluth potentials, 18, 131, 132
runaway electrons, 129–30, 136

sawteeth, 66
scaling law for confinement, 65–6
scattering of lower hybrid waves, 85–6
second harmonic heating, 46–52
singularity, Alfven wave equation,
    30–2, 35
slab geometry, 6
slide-away regime, 136, 142

slotted waveguides, 94
slow wave, 28–9
spectral gap, 84, 135
spectrum, Alfven wave, 31, 35–7
steepest descents, 12
Stix parameter, 60
stochastic heating, 80–3, 132
surface mode, 36

T-10, 116
TCA, 39–40
tearing mode stabilisation, 119
thermal velocity, 47
three-wave interaction, 85
TM-3, 116
tokamak, 1, 6, 42
transformer recharge, 137–8
transit time damping, 24–5
transmission coefficient, 14
trapping of electrons, 122, 142
two-ion hybrid resonance, 52–3

upper hybrid resonance, 99

variational principle, 49–50
Versator II, 136
Vlasov equation, 10, 12

wave energy flux, 6–7, 14
wave momentum, 23
Weber's equation, 14
WKB approximation, 6–7

X mode, 99
x-ray spectrum, 141